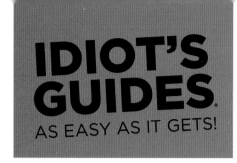
IDIOT'S GUIDES
AS EASY AS IT GETS!

Wine

by Stacy Slinkard

ALPHA

A member of Penguin Group (USA) Inc.

ALPHA BOOKS

Published by Penguin Group (USA) Inc.

Penguin Group (USA) Inc., 375 Hudson Street, New York, New York 10014, USA · Penguin Group (Canada), 90 Eglinton Avenue East, Suite 700, Toronto, Ontario M4P 2Y3, Canada (a division of Pearson Penguin Canada Inc.) · Penguin Books Ltd., 80 Strand, London WC2R 0RL, England · Penguin Ireland, 25 St. Stephen's Green, Dublin 2, Ireland (a division of Penguin Books Ltd.) · Penguin Group (Australia), 250 Camberwell Road, Camberwell, Victoria 3124, Australia (a division of Pearson Australia Group Pty. Ltd.) · Penguin Books India Pvt. Ltd., 11 Community Centre, Panchsheel Park, New Delhi—110 017, India · Penguin Group (NZ), 67 Apollo Drive, Rosedale, North Shore, Auckland 1311, New Zealand (a division of Pearson New Zealand Ltd.) · Penguin Books (South Africa) (Pty.) Ltd., 24 Sturdee Avenue, Rosebank, Johannesburg 2196, South Africa · Penguin Books Ltd., Registered Offices: 80 Strand, London WC2R 0RL, England

International Standard Book Number: 978-1-61564-416-2
Library of Congress Catalog Card Number: 2013935160

15   14   13   8 7 6 5 4 3 2 1

Interpretation of the printing code: The rightmost number of the first series of numbers is the year of the book's printing; the rightmost number of the second series of numbers is the number of the book's printing. For example, a printing code of 13-1 shows that the first printing occurred in 2013.

Note: This publication contains the opinions and ideas of its author. It is intended to provide helpful and informative material on the subject matter covered. It is sold with the understanding that the author and publisher are not engaged in rendering professional services in the book. If the reader requires personal assistance or advice, a competent professional should be consulted. The author and publisher specifically disclaim any responsibility for any liability, loss, or risk, personal or otherwise, which is incurred as a consequence, directly or indirectly, of the use and application of any of the contents of this book.

Most Alpha books are available at special quantity discounts for bulk purchases for sales promotions, premiums, fund-raising, or educational use. Special books, or book excerpts, can also be created to fit specific needs. For details, write: Special Markets, Alpha Books, 375 Hudson Street, New York, NY 10014.

Trademarks: All terms mentioned in this book that are known to be or are suspected of being trademarks or service marks have been appropriately capitalized. Alpha Books and Penguin Group (USA) Inc. cannot attest to the accuracy of this information. Use of a term in this book should not be regarded as affecting the validity of any trademark or service mark.

Publisher:
Mike Sanders

Executive Managing Editor:
Billy Fields

Senior Acquisitions Editor:
Tom Stevens

Editorial Supervisor:
Christy Wagner

Senior Production Editor:
Janette Lynn

Senior Designer:
William Thomas

Indexer:
Johnna VanHoose Dinse

CONTENTS

Wines of Europe 59

Wines of North America 125

Wines of the Southern Hemisphere 157

INTRODUCTION

Wine can be fascinating yet overwhelming, whether you're just starting to adventure into the world of wine or you've been sipping and sampling for a while. Even writing this book, which breaks down wine's complexities and nuances into basics beginners will understand, was intimidating endeavor for me—and I've studied wine for years.

Like any great field of study, the more you learn and understand, the more you realize how little you truly know. My pursuit of wine will always be an ever-moving target—quickly changing courses and diverting interests from well-known wine estates with familiar favorites to mastering the tongue-twisting pronunciation of a never-heard-of grape that's been grown in a remote region of Italy for centuries. My curiosity darts from the elements of making wine; to the pleasure of drinking wine; to the constant search for answers to the how, why, and where of buying, storing, and serving good wine with good food.

Each vintage tells its own story of how the growing season went and who handled it well from start to finish. With every sip, your palate develops an impression, a preference, and ultimately a profile that speaks to your individual taste. Wine provides a classical education—an intricate marriage of agriculture, science, geography, faith, family, language, art, and profound history that's reached new heights thanks to modern technology, communication, and global distribution. Wine cultivates its own relationships—welcoming visitors to taste the culture of a country (or county) and providing the warmth of food and friendship to its local neighbors. Wine communicates even when words cannot, spans cultures, leaps language barriers, and promises good times to those willing to share good food and good wine.

In this book, you get the fundamentals of wine—easily accessible answers to questions you might have about wine and some you haven't thought of. By taking a peek at how wine is made from grape to glass, and getting to know the world's most popular grape varieties in all their aromatic, delicious, food-pairing glory, you learn the easy way to select and buy wine for any occasion. You discover how to really taste wine, meet all the glassware and gadgets that go along with it, and learn about serving and temperatures and optimum wine storage conditions.

Then we take a tour of the world's most important wine regions and explore how they cultivate vines, label their wines, and pair their bottles with local foods. Keeping a constant eye out for wine in all its various expressions—red, white, or rosé; sparkling to sweet; fortified to late harvest and ice wines—the diversity of grapes, styles, forms, and fashion are virtually limitless with every region, vintage, vintner, and estate producing a different take on the grape year in and year out.

There's so much to discover! Cheers!

Acknowledgments

A special thanks to Dane for making it possible for me to work on this project in the first place by carving out the time and re-routing his work world; keeping humor and patience in high-gear; putting up with late, late nights; and always for his tenacious love. Many thanks to Taylor for her help with organizing images and glossary terms, and to McKinley for her culinary efforts, Caed for easy laughs, and Grant for constant countdowns. Thanks to Daryl and Robbie Slinkard for hosting kiddos so I could write, write, write. Huge hugs to Greg and Connie Taylor for watching, driving, and feeding my crew (chocolate-chip cookies included) while I was tethered to the computer. Thanks to Christy Fagerlin for making it possible to complete the work on Italy and Chuck Henry for reviewing my wine chemistry 101. To my family and friends who made it a point to check in on the progress of the project and offer sweet encouragement, thank you. My appreciation also goes to Thea Schlendorf for scouting high and low for various images. I'm also thankful for all the wine estates, winemakers, wine professionals, wine shop owners, fellow wine writers, and sommeliers who have been willing to share their passion for the vine and teach me so much through the years. Big thanks to Tom Stevens for asking me to write this book! Thanks also to Christy Wagner for her eagle eye, attention to detail, sense of humor, and patience and persistence with the editing process. For the tremendous effort in the graphics arena, my gratitude goes to Sarah Goguen.

Finally, I want to give a warm-hearted thank you to Alpha Books for giving me the opportunity to welcome those starting their venture into the world of wine with this stepping-stone.

WINE BASICS

Wine unites people and places, cuisines and cultures, faith and freedom, reflecting an expansive tapestry of spirited tastes, bold talents, and tenacious history. For being "just" fermented grape juice, a considerable amount of pomp, prestige, and enigma surrounds the favored fruit of the vine. And yet, there's an easy approachability, a worldwide welcome mat wine rolls out for friends, family, and friends-to-be.

Wine is made both in well-known and relatively unknown regions around the globe every year, with each vintage describing the weather, geography, and soil and whispering its own unique sense of place through its aromas and flavors. Thousands of grape varieties are grown around the world, with every bottle throwing the limelight on a single varietal or showcasing the synergy of a blend. From ground to glass, a grape's journey shares the story of its vineyard terroir, its distinct varietal contributions, and the winemaker's tale—the good, the bad, and the in between can each claim its share of the glass.

From how grapes are grown, to getting to know the various grape varieties (starting with the "big eight"), to buying wine for home or choosing a bottle at a restaurant, there's plenty to learn and taste along the way. When it comes to serving wine, some tools make a significant impact on the wine while others are just fun to use; I show you the difference. Tasting wine (preferably with food) is the end goal, and we get there while raising your sensory awareness so you can experience the most from the glass. Storing wine, whether for the short or long term, comes with caveats, and in the following chapters, you learn how to keep your wine safe from wine predators (oxidation, premature aging, and vibration). There are plenty of pieces in the wine buying, tasting, and storing puzzle, but once you become familiar with a few key concepts, the rest fall into place.

How Wine Is Made

Wine has been made for thousands of years, with some of the oldest evidence found on wine-stained artifacts from the Middle East dated more than 6,000 years old. Grapes and fermentation, the basis of how it's made now and how it was made way back then, hasn't changed all that much. Although the modern version lavishes considerable tradition and technology into the making mix, the short story of how wine is made goes something like this: grapes are grown, picked, crushed, pressed (prior to fermentation for white grapes and after fermentation for reds), fermented, clarified (or not), aged in oak or bottled, and sold.

However, every step along the way is teeming with choices. Which grapes should be grown and where for solid sun exposure and good soil drainage? When should they be harvested? The weather and the grape's sugar levels are integrated determinants of harvest time decisions. Are they red grapes or white grapes? The answer will determine if the grapes are fermented with or without the skins on. Should fermentation take place in stainless-steel tanks or oak barrels? What about aging? Should the wine age in oak, in the bottle, or both?

The way a winemaker answers these and many more questions ultimately determines the style, quality, and quantity of wine made.

GROWING GRAPES AND THE HARVEST

The majority of the world's wines are made from the *Vitis vinifera* species of grape, which brings extraordinary variety—including all the well-known international grapes—to the winemaker's bottle.

The age-old question of where to grow grapes has raised awareness of and placed a priority on the increasing importance of terroir. *Terroir,* loosely translated as "soil" or "region" in French, covers all aspects and potential influences of the specific soil structure, topography, irrigation, drainage, sunshine, temperatures, climate, weather patterns, and general geography of a grapevine's immediate growing area. Different grapes thrive in different regions. Some are happiest in cooler climates, like Riesling and Germany, while others reach incredible levels of ripeness in sun-soaked regions, like Shiraz in South Australia. Where a grape is planted and when it's harvested is largely determined by terroir.

Harvesting grapes is based on the maturity of the fruit and the specific levels of sugar (called Brix) and acidity within the grape. In general, the more sun grapes get during the season, the riper and more flavorful the grapes are at harvest. Riper grapes hold higher levels of sugar, which ultimately translates into higher alcohol levels and typically a fuller-bodied wine. That's one reason why German wines tend to run a little leaner on alcohol levels and why sunny California can make some fairly big wines with higher alcohol levels and very full bodies.

Grapes are harvested either by hand or machine, depending on the region. The picked grapes are brought to the staging area called the crushpad. Typically, they run through the destemmer-crusher machine next. There are two parts to the crush process. First, the grapes go through the destemmer to separate the grapes from the stems. Then, usually the grapes are crushed—traditionally by foot but today by mechanics, often with a machine aptly named the crusher that splits the grape's skin, releasing the juice and pulp.

After making their way through the destemmer-crusher, the "crushed" grapes are sent either to the press (for white wines) or the fermentation tanks (for reds). The press squeezes more juice from the white wine grapes. The skins are retained while the juice goes into the fermentation tanks. With red wines, the juice and the skins go into fermentation to soak together. This process—when the red grapes are soaked in the juice to extract color, flavor, tannin, and aromatic intensity—is called maceration.

Red wines are crafted from black-skinned grapes, and white wines are made from green-skinned grapes.

Rosé wines are another story. Most often, they're created from red-skinned grapes that have only spent a brief time in contact with their grape skins. On occasion, they're made from white wines with a splash of color from red wine contributions.

Sparkling wines are produced from *still wines* that can either be based on red wine or white wine grapes that go through a second closed fermentation. This keeps the bubbles (CO_2) trapped inside. That's what gives these wines their sparkle.

Fortified wines are still wines made from either white wine or red wine grapes that have additional alcohol, often brandy, mixed in.

All red, white, rosé wines that lack any kind of bubble component are called **still wines.**

FERMENTATION

The process of turning grape juice into one of the world's most coveted elixirs is called *fermentation*. In wine-making, fermentation is where the magic happens. It's where yeast, either naturally occurring or intentionally added by the winemaker during "inoculation," is added to the grape juice. The yeast feeds on the sugar (glucose and fructose) in the juice, converting it to alcohol (ethanol) and carbon dioxide (CO_2). The more sugar that's metabolized into alcohol, the higher the alcohol content, and the drier the wine. When the yeast's conversion of sugar to alcohol is stopped before all the sugar is turned into alcohol, the result is a sweeter wine.

Fermentation creates a number of fragrant chemical compounds as the yeast methodically eats the sugar and releases alcohol. These highly aromatic compounds often smell like a medley of familiar fruits, spices, flowers, or even butter (in the case of malolactic fermentation). They're the reason why wine—made only from grapes—can smell like violets, cherry, cocoa, pineapple, grass, and more.

Fermentation typically takes place in stainless-steel tanks (used mostly for white wines), concrete vats, oak barrels, and even the bottle itself (as is the case with sparkling wine production). The whole process of fermentation can take anywhere from days to months, depending on yeast, temperature, and a host of other conditions.

OAK'S INFLUENCE

When wine comes into contact with oak, either during fermentation or aging (or both), the wine is never the same. Oak concentrates a wine's colors, amplifies aromas, adds significant complexity and underlying richness, and creates a creamy texture. Oak's hallmark aromas include vanilla, cinnamon, clove, caramel, butter, butterscotch, coconut, wood, cedar, smoky nuances, and sweet toasted or caramelized notes. The aromas are the first serious sign of oak's influence, and these same scents can echo along the palate as flavors and again in the form of rich, creamy textures and a slight increase in tannins. Wine may spend several months to several years aging in oak barrels before being blended and bottled, depending on the style and end goal of the wine.

Oak barrels are made in a variety of sizes, with the standard being a 60-gallon (225-liter) barrique. Although larger barrels have more surface area for the wine to make contact with, they also hold a larger volume of wine, and the majority of that wine does not touch the oak nor reap its influence. So when oak character is desired, the smaller, 60-gallon (225-liter) barrels have become the most common for wine making in Europe and the New World.

Oak for the barrels is sourced primarily from forests in France (the most expensive), the United States, and Eastern Europe, and each region's oak barrel is known for imparting its own unique aromas and flavors to the wine. The newer the oak barrel, the more influence it has on the wine. As a barrel ages, the impact the oak has on the wine decreases, and within several vintages of use, the barrel may well be deemed a "neutral" barrel—essentially a holding tank void of notable flavor, aroma, and tannin contributions.

Red wines generally see a lot more oak than white wines, with the exception being Chardonnay. When it's well oaked, Chardonnay takes on darker colors; a more butter-driven character; and rich, creamy, round textures. Wine labels use several terms (beyond the flavor descriptors of *vanilla, spice,* and other signature smells) to indicate the wine has befriended oak in the wine-making process: *oak aged* is the most obvious, but there can also be bits about the barrel itself like *barrel fermented, barrel aged,* or *barrel select.*

Another point about oak: it's pricey. So to skirt the price gauging, winemakers might opt to add "oakiness" in other ways. Oak staves (like sticks), oak chips, and even oak extracts may be added to wine to boost the oak character without boosting the price. When winemakers use the real deal and buy barrels, the cost is passed on to the consumer. This is one reason (among many) wines higher on the quality ladder typically cost more. Some wineries keep costs down by using a fairly common oak compromise, where they age a percentage of wine in new oak and the rest in neutral barrels and blend the wine before bottling. This ensures traditional oak-aging practices without the price tag.

Toasting is another factor in the oak equation that plays a significant role in what the wine picks up from the barrel. Barrels are crafted over small open fires to make the wood pliable so it can be arched into the traditional curved shape. During this heating process, the inside of the barrel is charred or toasted. The level of toasting varies, and barrels can be light, medium, or heavy toasted. Winemakers pick the type of toasting they prefer for their wines, much like a chef selects seasoning ingredients for a signature dish.

OLD WORLD VERSUS NEW WORLD WINES

Wine-growing regions are places of distinction, with unique soil structures, climate zones, geographic features, and grape-growing abilities. A particular wine-growing region can be assessed in terms of its broader geography, whether it's from the Old World—meaning the traditional wine-growing countries of France, Italy, Germany, Austria, Spain, and Portugal—or the relatively "new" New World—North America, South America (predominately Argentina and Chile), Australia, New Zealand, and South Africa.

Old World wine regions focus on the *place*. You see it most obviously on wine labels, with places like France, Italy, Germany, and Spain putting added emphasis on *where* the grape was grown not *which* grape was grown. That's why in the famous French wine-growing region or appellation of Burgundy (*Bourgogne* in French), labels feature important vineyard villages or communes like Chassagne-Montrachet, not Chardonnay or Pinot Noir (the grapes grown).

The concept of terroir—literally the "soil or region"—encompasses everything from the dirt the vines are cultivated in, the geography of the region at large, and the hillside topography of the specific vineyard, to the amount of sun a place does or doesn't receive, irrigation and drainage issues, and the impact of climate and weather patterns. All these factors combine to give the grapes a sense of place, with each vine or vineyard reflecting that place and emphasizing different attributes ultimately expressed in the wine itself.

New World wines give priority to which grapes are grown in a given region. Technology tends to take greater influence than tradition in New World regions, with wine-making practices enjoying the added benefit of recent research, high-tech machinery, and soil and grape testing capabilities unheard of several decades ago.

Old World regions tend to be located in colder climates. Vintage years can be more of a factor, with the cooler climes often producing wines with more subtle fruit and greater acidity that can spotlight earthy, mineral notes and elevated finesse as a reflection of the soil and climate. New World wine regions by and large enjoy warmer weather with solid sunshine, yielding fruit with greater levels of maturity and ripeness. That often means fuller-bodied wines with forward-fruit flavors and higher levels of alcohol. These wines might also come in a little lower overall on the acidity scale.

New World wines can be fairly extroverted, with big, bold flavors, and Old World wines frequently show more subtle, elegant styles.

EIGHT GRAPES TO KNOW

Each year, thousands of unique grape varieties are planted, picked, pressed, and fermented to produce more than 30 billion bottles of wine. That's a staggering number, especially if you're just venturing into the world of wine. Where do you start? How do you choose?

Many enthusiastic buyers pick wines based on pretty labels, while others choose a favorite grape and stick with it. Some approach the wine market like a spreadsheet—systematically covering wine by grape variety, renowned region, or popular food-pairing principles. Yet many would-be wine fans simply go by "professional" word of mouth or the latest wine critic's recommendation, often missing cues from their own palate preferences in the process.

When it comes to understanding wine in all its glorious forms and fashions, knowing the look, smell, taste, and general style of eight grapes—four white and four red—gives you a strong starting point. Riesling, Pinot Grigio (also tagged Pinot Gris), Sauvignon Blanc, and Chardonnay cover the initial white wines to know (in order of light to full-bodied profiles). Pinot Noir, Merlot, Cabernet Sauvignon, and Syrah (also called Shiraz Down Under) represent the key players in the reds (again in order of lightest to fullest body). These eight grapes offer infinite insight into the world's major international varieties and provide glimpses into classic styles as well as the influence of regional terroir.

By tasting these wines in side-by-side comparisons, you learn to detect subtle and striking nuances in terms of color, aromas, flavor, and overall style. By sampling the range of white or red wines in order of light-bodied to full-bodied profiles, you can compare and contrast colors, aromatic distinctions, and taste sensations and ultimately deem a grape one you like or not. Likewise, trying one of the eight grapes from two or three different regions brings another illuminating angle on a particular grape variety. Tasting a Riesling from Washington State and an Old World Riesling from its original German roots, for example, highlights the differences in each grape's growing environment, vintage conditions, and wine-making style.

Riesling, Pinot Grigio, Sauvignon Blanc, Chardonnay, Pinot Noir, Merlot, Cabernet Sauvignon, and Syrah are popular international grapes with winemakers and wine buyers because they offer a wide range of flavor, body, and food-pairing compatibility. They're also consistent, reliable grapes, producing outstanding wine all over the world.

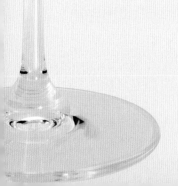

RIESLING

Typically grown in cool-weather climates, with Germany being Riesling's original home sweet home, this delicate white wine grape has also shown its inherent versatility in Alsace, France; Austria; and Australia; as well as New York, Washington, and California in the United States. Traditional German Rieslings tend toward lower alcohol levels with a splash of sweet, especially in the Spätleses and Ausleses. In France, Alsace loves to show off Riesling's racy acidity in a bone-dry, fuller-bodied style. In North America, New York Rieslings often shine as ice wines. Australia's intense sunshine ushers in exotic, tropical fruit character to many of it's Rieslings.

Often saddled with the incomplete reputation of being a "sweet" white wine, Riesling also presents as off-dry and even completely dry. Although *sweet* seems to be an easy moniker for Riesling, *acidity* would be far more accurate. The grape's natural acidity keeps the wine's innate levels of sugar in check and serves as the fresh, food-friendly accompaniment to many demanding dishes. Without the lively acidity, Riesling would come across as flat and flabby on the palate.

In the Glass

Rieslings range from practically clear to pale yellow with a green tinge. The delicate floral aromas and flavor impressions are accompanied by crisp and refreshing green apple, zesty citrus, plush peach, or other tropical fruits. Riesling also is capable of reflecting a soil-driven flintlike minerality, especially in German and Alsatian bottles.

On the palate, Riesling charms with undeniable fruit character and offers a range of light to medium-bodied profiles. It can quickly turn toward full-bodied, however, with robust, syruplike textures in the grape's dessert wine categories.

With Food

Riesling could easily be voted the most food-friendly white around. Off-dry Rieslings take to everything from spicy curries to shredded pork or fish tacos; from Cajun to Asian dishes; all the way to shellfish, fried fish, and smoked fish. Roasted pork, poultry, and veggies, along with hard-to-match salads with acidic dressings and a tremendous array of appetizers all play well with the slightly sweet styles of Riesling. Save the fuller-bodied, honeyed flavors of dessert Rieslings for fruit-filled desserts, puddings, crème brûlée, caramel- and toffee-themed treats, as well as cheesecake, assorted cookies, and vanilla-based cakes.

PINOT GRIGIO

Pinot Grigio in Italy, *Pinot Gris* (the *s* is silent) in France, or *Ruländer* in Germany—it's all the same dry, white wine grape under different regional names, but it typically brings a light to medium-bodied profile, often with an easygoing sipping spirit and a fairly affordable price tag. The terms *grigio* and *gris* are translated as "gray," giving reference to the color of the grape's skin.

Italy tends to be the grape's first point of reference, with the country's best versions coming from Alto Adige, but the majority of Italy's Pinot Grigio can be a fairly simple sip, offering subtle aromas with a straightforward palate appeal. Pinot Grigio tends to be more of an introvert than extrovert, not wanting to rock the boat, flaunt its flavors, or make you sit up and take excessive notice. This is a low-key, pretty laid-back wine.

In France, however—specifically in the northeastern region of Alsace—the grape soars to new heights under the moniker of Pinot Gris. Alsace turns things up a notch and coaxes the grape into brighter fruit flavors, greater weight and texture, and more compelling complexity.

In the United States, Oregon often pulls out the grape's rich fruit character, spotlighting more depth and adding an extra splurge of acidity in its cooler microclimates.

In the Glass

Clear to straw yellow in color, Pinot Grigio and Pinot Gris scents range from fairly subtle to the aromas of almonds, citrus, apple, and peach, and at times a touch of minerality, especially from Alsatian versions. On the palate, the spectrum spans from lighter-bodied with an emphasis on clean and crisp to the medium-bodied impressions showing more forward fruit, richer textures, and greater complexity.

The cooler the climate, the better the grape seems to express itself. Naturally weighing in a little lower in acidity, cool growing regions help perk up the acid levels and allow for better backbone and sense of structure.

With Food

A fairly flexible grape, thanks to its delicate nature, often subtle nuances, and lower levels of alcohol, Pinot Grigio partners well with lighter lunch fare. Consider it alongside a variety of creamy salads, shellfish and lightly seasoned fish, light pastas, grilled chicken, or BLT sandwiches.

SAUVIGNON BLANC

Sauvignon Blanc's Old World roots are firmly planted in the cooler climate of France's Loire Valley, where it goes most often by its regional place names of Sancerre and Pouilly-Fumé. Subtle and refined, with a crisp, mineral-laden backdrop, Sauvignon Blanc exudes classic French elegance. Yet it remains fairly feisty, with fruit that rocks from Granny Smith apple to zesty lime and full-on grapefruit all the way to the plush flavors of peach and honeydew melon when grown in the Sauvignon Blanc hot spots of the United States, Chile, and various warm-zone microclimates of New Zealand. But whether in France, the States, or the Southern Hemisphere, Sauvignon Blanc often brings aromatics with a distinct pungency, reminiscent of herbs and fresh-cut grass. Sauvignon Blanc is tasty, tangy, and tart; pleasant and piercing; yet incredibly capable of handling "hard-to-pair" foods with lively style and abundant acidity.

Sauvignon Blanc is also labeled Fumé Blanc in the United States—same grape, just a different name. You'll also see Sauvignon Blanc carrying a bit more fruit and oak in the States, where warm-weather growing conditions can result in a fuller-body profile spotlighting generous, sometimes tropical fruit aromas and palate impressions. Offering great value, good accessibility, and a stark departure from the polar profiles of Riesling and Chardonnay, Sauvignon Blanc is delicious and different.

In the Glass

Ranging in color from pale yellow in France and New Zealand to more golds and gloss as a result of optional oak in the United States, Sauvignon Blanc creates a full medley of aromas, with an emphasis on orchard fruit, citrus, and herbal undertones. These fabulous fruit smells are typically reflected on the palate (minus the fresh-cut grass). Most often found with a dry, medium body; fantastic acidity; and a crisp, clean finish, Sauvignon Blanc runs a little leaner in cool climate zones.

With Food

Providing extraordinary pairings with a variety of fresh herb–inspired dishes, roasted seasonal veggies, pungent goat cheese (stuffed in sweet red peppers and drizzled with olive oil and balsamic vinegar), as well as a slew of shellfish recipes, well-seasoned chicken dishes, and creative vegetarian fare (including the tough-to-pair duo of asparagus and artichokes), Sauvignon Blanc brings some serious synergy to wine and food pairing.

CHARDONNAY

Chardonnay reigns as the world's most popular white wine grape, growing successfully in virtually every major wine-producing nation, thanks to the grape's ability to adapt well to a variety of climates and conditions. In Burgundy, France, Chardonnay's hallowed homeland, the wine made from Chardonnay grapes is referred to as white Burgundy, Pouilly-Fuissé, or Chablis (the latter two are regional names known for growing great Chardonnay). Setting the gold standard for Chardonnay around the world, Burgundy's cool climate zones showcase crisp, pure fruit character centered on the softer scents and tastes of ripe pear; fresh apple; and an earth-driven, chalk and flint mineral component.

In warmer growing regions like California and Australia, Chardonnay grapes tend to reach greater degrees of ripeness, with richer fruit aromas and flavors that venture toward apricot, peach, melon, and mango. These wines also present fuller bodies with higher alcohol levels.

In the Glass

Chardonnay spans the color spectrum from straw yellow to golden yellow, with oak, age, and warm weather bringing out the golds. Dry and often full bodied, Chardonnay takes well to oak's enduring influence. Yet the direct relationship between oak and Chardonnay has come under considerable scrutiny, as many bulk-bottled New World Chardonnays have been the victim of "overoaking," yielding wines that bust out of the bottle with big, buttery character and limited fruit. Oak should accent the fruit, not completely cover it. As the pendulum swings back to center, stunning expressions of Chardonnay, with an exquisite balance between the fruit and oak, are easier to access. In some instances, producers intent on pushing the pendulum in the opposite direction are hitting the shelves with a surge of "unoaked" Chardonnay, exposing the grape's unobstructed, pure fruit profile.

With Food

Chardonnay's fuller body gravitates toward heavier fare, and recipes that call for cream and butter make a delicious match with Chardonnay that's seen a bit of oak. Think lobster drenched in butter or swirling in bisque, crab cakes, roasted or baked poultry dishes, soft and creamy cheese spreads, Alfredo-themed pastas, and fresh shellfish.

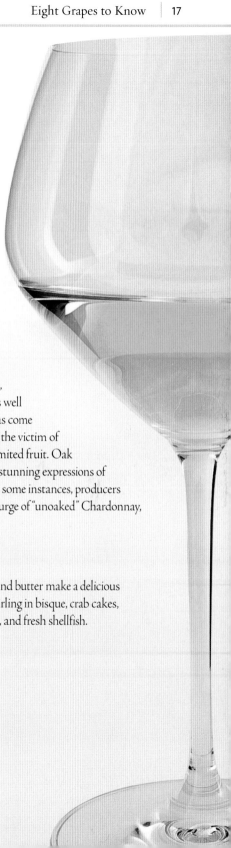

PINOT NOIR

Pinot Noir's prestigious roots run deep in Burgundy, where it's often called red Burgundy and exhibits an unmistakable earthy character, with refined fruit, silky textures, and a pliable profile. New World renditions of Pinot Noir (often shortened to just Pinot, with a silent *t*) run heavier on the bright red berry front, often with raspberry, strawberry, and cherry taking center stage. The tannins are notably lower in Pinot, which really allows the signature silky texture and delicate structure to shine.

Pinot Noir is a fairly finicky red grape to grow. It's thin-skinned, often unreliable, yet highly demanding, requiring cooler climates and presenting a litany of challenges to get it "just right."

Pinot Noir producers are paramount, especially in Burgundy, where the vineyards have been sliced and diced over the decades to bring multiple proprietors into the ownership mix. Vintage also plays a key role in determining the quality and ultimate expression of the grape in a given year.

In the Glass

With its ruby red colors and aromas ranging from fresh flowers, to earthy mushroom (or "forest floor"), to the best of the berries, to a hint of sweet vanilla oak influence, Pinot Noir takes care to keep tannins in check, thanks to the grape's thin skin. It typically features a light to medium-bodied style with cool climate–infused acidity. Textures are sometimes reminiscent of white wines, given the softer palate appeal, delicate body, lively acidity, lighter tannins, and forward fruit. In fact, many white wine lovers find Pinot Noir an easy entry point into the often-daunting world of red wine.

With Food

Pinot Noir is to the red wine world what Riesling is to the white wine world. It's *the* go-to wine for everything from international cuisines to down-home favorites. Extremely versatile, befriending everything from fish sticks, to herb-crusted salmon steaks or ahi tuna, to roasted poultry, to beef topped with Burgundian-themed mushroom sauce and even burgers and brats, Pinot Noir's ease with food makes it a top red wine pick for pairing with a range of dishes.

CABERNET SAUVIGNON

Cabernet Sauvignon, often shortened to just Cab, is another international red grape variety that boasts plenty of flexibility when it comes to where and how it's cultivated. Bordeaux, France, is its original hometown, but Cabernet Sauvignon also thrives in a number of warm-weather regions, with California, Washington, Chile, and Australia leading the pack. In Bordeaux, Cabernet Sauvignon's tannins are typically tamed by blending with Merlot and Cabernet Franc. The best blends produce wines that focus on fruit—*blackcurrant* is a classic description, although plum, black cherry, and oaky mocha themes are also seen. At the same time, the blends manage to highlight the fruit's soil, minerality, and oak-driven components. Bordeaux is about balanced blends.

Preferring warmer regions in general, New World Cabs are completely capable of expressing their forward-fruit profiles with ripe, jammy fruit flavors that include plum, blackberry, dark cherry, and blueberry; dark chocolate or espresso notes; and oak's most obvious influences, smoke and heady vanilla spice.

In the Glass

Maroon red with deep garnet color schemes, with scents ranging from concentrated dark fruit aromas to wood-centered, earth-driven character and a decidedly dry, full-bodied style, Cabernet Sauvignon can pack a powerful punch on the palate. With understated fruit and traditional Cab class from Bordeaux, or gregarious and full of fruit, substantial tannins, and lots of power from its second home in California, this grape runs the gamut in terms of style and structure while offering considerable aging potential in the higher-end wines.

With Food

Best with beef, Cabernet Sauvignon is often a meat-lover's favorite. Grilled steak, roasted rack of lamb, hickory-smoked baby back ribs, pork roast, peppery sausage, prime rib, wild game, and aged cheeses all make suitable partners. At dessert, Cabs pair well with the tougher tannins in dark chocolate.

MERLOT

Dominating Bordeaux's vineyards for decades, Merlot grapes produce a dry, medium- to full-bodied red wine with higher levels of alcohol and often (but not always) softer on the tannins than its esteemed blending partner, Cabernet Sauvignon. The Merlot grape typically thrives in the warmer climes of Washington, California, and Chile.

Similar on many fronts to Cabernet Sauvignon, Merlot can come across a little softer, a little fruitier, and a bit more laid back than the more intense palate personality often associated with Cabs. Merlot is bottled most often in Bordeaux as a blend with Cabernet Sauvignon and Cabernet Franc. The most famous bottling is undoubtedly from Pomerol's Chateau Petrus on Bordeaux's distinguished right bank, which is made with 100 percent Merlot and comes with a shockingly high price tag.

New World Merlot is most often seen as a single-varietal wine, but blends are becoming more common. Merlot plantings reached their zenith in the United States in the mid-1990s as consumers sought softer reds, although quality took a beating as quantity became the focus. Recent years have seen a renewed commitment to maintaining Merlot plantings in the right soil structures (which is often but not always clay based) and microclimates to coax mature, well-ripened fruit to shine.

In the Glass

Ruby red to dense purple in color, Merlot offers a range of elegant, understated fruit with a bevy of complex aromas in Old World wines, and plenty of ripe fruit, round flavors, and warm spice shine in New World editions. From plum to cherry with nuances of dusty cocoa and tobacco leaves, Merlot boasts ample body, fine-grained tannins, and velvety textures. The wine's overall structure and palate profile often mirror Cabernet Sauvignon, with perhaps a little less tannic punch.

With Food

Merlot is made for meat. Hamburgers, grilled sausage, steak, Bolognese, lamb chops, pork chops, and even turkey and roasted duck are all worthy partners for Merlot. Strong, flavorful cheeses like Stilton, Roquefort, cheddar, Jarlsberg, and Camembert also make their mark on Merlot.

SYRAH/SHIRAZ

Running along the Rhône River in southwest France, the northern Rhône Valley has first dibs on the gutsy Syrah grape, which is labeled Hermitage, Côte-Rôtie, or Crozes-Hermitage for the village names, not the grape, in true Old World naming style.

Australia, however, takes this red wine grape—tagged locally as Shiraz—and flips the fruit forward into an undeniably, jam-driven fruit extravaganza balanced out by smoky black pepper and pumpkin pie spice with malleable tannins. Well labeled, well priced, and very well received, Australian Shiraz has made this lively yet concentrated version of a classic international grape available to fans worldwide.

Washington, California, and South Africa have also given this grape a fresh, fruity face with warm-weather growing factors kicking into high gear to produce well-ripened fruit.

In the Glass

Inky and *purple* are classic descriptors used when trying to capture Syrah's heavily pigmented color components. Rich, ripe fruit with aromas that stick to earthy, almost herbal undertones wrapped in warm, engaging spice, smoke, and black pepper give way to bright, sweet fruit profiles. With easy diversions toward cherry and chocolate, or tobacco and tar, along with leather and black licorice, you'll find plenty of palate intrigue with Syrah. The distinctly dry red wine maintains moderate to high tannins with a robust body and often comes in a little higher on the alcohol scale.

With Food

Perfect for pizza night, barbecued beef or chicken, smoked game, and pork or poultry, the ripe, forward fruit; velvety textures; and immensely approachable style of Australian Shiraz makes it a natural for pairing with all sorts of meaty fare. The more refined, traditional take on Syrah from the Rhône Valley makes a tasty pairing with black pepper steak, wild game (especially venison), and more mature cheeses.

BUYING, SERVING, TASTING, AND STORING WINE

The process of buying wine can be a joy or a nightmare, depending on where you're buying, what you're buying, and how you're deciding on the wine selection. By keeping a firm focus on cost parameters, personal palate preferences, food-pairing versatility, and whether the wine is for now or later, you'll build a wine-buying framework that will serve you well, whether you're selecting a wine for dinner tonight, dining out with colleagues, or shopping for an upcoming celebration.

After you've chosen a wine, it's time to pour. Bottle-opening technique, glassware considerations, decanting (or not), and serving temperatures all factor in to the wine-tasting experience.

Wine tasting differs from just drinking a glass of wine. It slows down the process to allow firsthand interactions and impressions. By exploring a wine's broad spectrum of color components, innate aromas, and total taste perceptions, you'll be equipped to determine its balance, food-pairing synergy, and overall style. This is where genuine, individual wine evaluation begins and where the stage is set to form lifelong favorites.

In the following chapters, I lead you through the fascinating process of wine tasting and explain how wine made solely from grapes can smell and taste like other familiar fruits. Then we take a look at how to store, where to store, and what to store, so you'll be prepared to extend the life of your wine for future enjoyment.

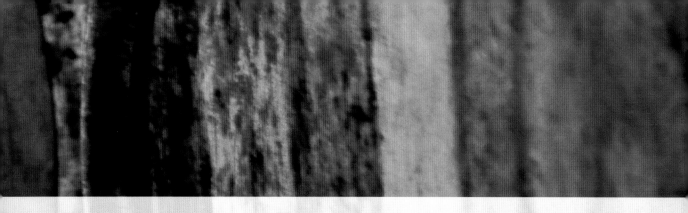

BUYING WINE

Buying wine today means tremendous selection, both in terms of wine types and wine shops. Whether it's at a local specialty wine shop or an online wine merchant, from a big-box store or a neighborhood grocery store, through the winery itself or via a reputable wine club, many compelling and convenient avenues are available for purchasing bottles (or cases) of wine.

Consumer access to international wines and regions has never been better. Online wine sales continue to soar, even with perplexing interstate wine shipping restrictions and numerous regulatory hurdles. Internet wine retailers and various smartphone wine apps offer enthusiastic shoppers the ability to sift quickly through top wine picks with search features and filters that scout for wines based on grape type, regions, style, vintage, price, ratings, points, and producer.

Wine clubs offer another route for adventurous wine buyers. It's easy to slip into the same wine-buying routine, or run with similar regional wine themes, but having a savvy wine club select and ship wine to your front door on a regular basis forces you to expand your wine horizons; broaden your palate; and typically learn, taste, and pair something new with dinner as a delicious result.

Big-box warehouse wines typically win in the value-pricin g category, scoring solid discounts when buying in significant bulk, but they can fall short in selection. The opposite tends to hold true for specialty wine shops, which seek out strong ties with boutique wineries and smaller growers, ordering hard-to-find wines, and often delivering extensive consumer education in the process.

When buying wine, keep the four C's of cost, choice, compatibility (with food), and cellar potential in mind as you peruse wine shelves, pursue magazine recommendations, or follow word-of-mouth wine-buying tips.

THE FOUR C'S OF BUYING WINE

When it comes to wine-buying strategies, you have several factors to consider—namely, the four C's: the cost of the wine, choice of wine styles, compatibility with food, and cellar potential.

Ask yourself how much you're willing to spend on a bottle, and think about what kinds of wine styles you typically enjoy. When it comes to food, what are your favorite bites, and how will the wine support those recipes? What about aging potential? Are you looking for wines you can buy at lower price points and then store for a bit, or will your average wine storage needs be the time it takes you to travel from the wine shop to the table? Let's look at each of these four C's in a little more detail.

Wine Cost Cutters: New Places, New Grapes, Second Labels

Determining cost upfront immediately narrows your focus when buying wine. How much or how little would you like to spend on a bottle? Is it an everyday wine for dinner tonight, or is it a special-occasion wine intended to mark a milestone or celebrate a holiday dinner?

If times are tight and you're hoping to lean more heavily into bargain bottles, scout for wines from up-and-coming or lesser-known wine regions. Places like New Zealand, Chile, Argentina, Portugal, and even South Africa have a history of wine production, but the last decade or two has seen a real surge in quality and export access, with many reasonably priced wines eager to extend tasty introductions.

Also keep an eye out for unusual grape varieties. Venturing out from under the umbrella of familiar international grapes and into local loves like Austria's crisp, dry Grüner Veltliner; the cool, clean lines of Italy's Soave; the spicy style of Chile's Carmenère grape; South Africa's smoky Pinotage; California's flavorful red Zinfandel; Spain's sultry Rioja reds—the list goes on, but you get the point. If it's new to you, give it a go. Not only will you expand your palate education, but you'll likely score some serious savings, too.

Many prominent wineries that have built a stellar reputation on high-end wines also bottle less-expensive second labels, or second wines, typically named with a nod toward the original estate. Famous examples include the super-pricey Bordeaux First Growths like Château Latour bottling under the second label of Les Forts de Latour or Château Lafite Rothschild's second label bottled under Carruades de Lafite Rothschild. Château Palmer tags a second label under Alter Ego. Cult California Cabernet Sauvignon producers also follow the second label trend, with Screaming Eagle running a second label named Second Flight, and Harlan Estate's The Maiden and Opus One also offering Overture.

Many second labels are impressive, and although they're often only a shadow of the world-renowned wines themselves, they still showcase context and glimpses of house style and are often made with input by the same wine-making team that forges the big-league bottles.

So Many Choices, So Little Time

When it comes to choosing a wine to buy, past preferences and personal palate appeal often hold the most sway. When you find yourself in a wine rut, it's a terrific time to evaluate what you like about a certain wine style and then springboard off in a new direction while reaching for similar vinous attributes.

Do you like fruit-forward, low-tannin red wines? Then Beaujolais, Spain's Rioja Crianza, or Pinot Noir may serve you well.

Prefer big, bold, burly red wines? Hunt for Italy's Barolo or Brunellos, Spain's Priorato, California Cabernet Sauvignon, or a variety of Chilean or Argentinean red blends.

Dry, crisp, white wines more your style? Opt for the vibrant, clean lines of Sauvignon Blanc or Sancerre, the lively character of Grüner Veltliner, or the fresh finesse packed into Burgundy's Pouilly-Fuissé.

Sweet dessert wines a particular passion? Check out Portugal's long-standing dessert darling, Port, or other fortified finds like Madeira (also from Portugal), Banyuls from southern France, Italy's Marsala, Spain's Pedro Ximénez Sherry, or late harvest delights from around the globe, with particular (although pricey) attention to Germany's Beerenauslese or Trockenbeerenauslese, as well as ice wines.

Complete Compatibility: Food and Wine Pairing Practice

Wine and food have gone hand in hand for centuries, each bringing out the very best in the other. So it stands to reason that dominant dishes, favorite family fare, and regular recipes should impact wine-purchasing plans.

Do you steer clear of red meat and bring more vegetarian-based meals to your table? If so, opt for lively white wines.

Are hearty Italian influences with fresh tomatoes, herbs, beef, broth, sausages, and al dente pastas weekly staples? Local reds like Chianti, Primitivo, Barbera, and Dolcetto should grace your glass.

Everyday table wines are eager to support regular menu plans, so to get the most bang for your food-and-wine-pairing buck, grab several "go-to" wines with a versatile pairing nature and plenty of acidity.

Cellar Potential

Simple, everyday, ultra-affordable wines are intended to be enjoyed sooner rather than later. Contrary to popular thought, the vast majority of today's wines are not built to age. In fact, a very small percentage of today's wine market actually improves with extended time in the cellar.

Most wines purchased for their cellaring potential are bought upon release, when prices tend to be lowest. With that in mind, if you're interested in buying wines to cellar, look toward the big boys—Bordeaux (including Sauternes); high-end Burgundy; Italy's Barbaresco, Barolo, and Brunello di Montalcino; some German Rieslings (thanks to the high acidity) and some dessert wines; along with vintage Port, California's Cabernet Sauvignon (and Cab-based blends), Rhône Valley Hermitage and Côte-Rôtie, and high-end Australian Shiraz.

Heavy tannins in red wines and high acidity in white wines are big indicators of aging potential because proper cellaring tames the tannins and mellows the acidic edge.

WINING AND DINING

Good wine restaurants aren't always the ones with the most exhaustive wine lists. Rather, they're ones with a passion for pursuing engaging wines at various price points. These outlets take pride in serving both local and international wines side by side with an eye toward value and food-pairing compatibility.

In an industry known for dramatic wine mark-ups, it's always refreshing to find restaurants that keep wine prices reasonable and place priority on updating inventory. If the restaurant staff knows and understands the layout of the wine list and how the wines should play with the menu items, you're in luck.

My favorite restaurants are those that take the time and genuine interest in ensuring their patrons are as comfortable as possible with both the wine list and the menu by making wine-pairing suggestions on the main menu with each entrée choice. These are often the types of venues that are just as committed to server education as they are to providing top wine selections. Feel free to ask questions, gather input, or seek specific pairing recommendations. Dining out at wine-conscious restaurants is just another avenue for adding to your ongoing wine education.

Navigating the Wine List

Wine lists are arranged a variety of ways. Often it's by the color or style of wine—red, white, rosé, sparkling, or dessert; or by body—light, medium, or full bodied. Other lists are organized by regional scope, giving attention to wines by country of origin or in broader terms like Old World wines versus New World wine themes. Categorizing wines by flavor tips like dry, crisp, refreshing whites or big, bold, full-bodied reds tends to offer customers more information for the big ordering decision.

If you're looking to broaden your wine résumé (and save a few bucks), keep an eye out for the less-familiar grapes from smaller growing regions. An Italian Soave, although not as popular as Chianti, offers another view of Italy's grapes and may even run a little cheaper.

Wine lists also print the price of the wine, often broken out by cost per glass or half glass and full bottle or half bottle. Take advantage of menus that list wines by the glass to experiment with different wines easily and without the commitment of buying a whole bottle. Buying an unfamiliar bottle might seem like a risky way to spend your wine bucks (retail or restaurant), but paying by the glass can feel less frivolous.

By the Bottle or by the Glass?

Depending on how much wine you and your dinner party plan on drinking over the course of your meal, it might make more sense to buy a bottle of wine instead of buying individual glasses, especially if you order more than three glasses.

When buying a bottle, you should consider what appetizers and entrées are being ordered. What do your fellow diners prefer—rich, spicy, dry, sweet, light, or heavy wine? Are they eating spicy dishes, meat and potatoes, seafood, or more exotic cuisine? Which wines can handle the most diversity with the dinner menu?

In general, the easygoing styles of many fresh-faced red wines like Beaujolais, Rioja reds, Pinot Noir, and Aussie Shiraz are easy pairing partners for a tremendous variety of foods. Riesling, Sauvignon Blanc, Grüner Veltliner, and dry-styled rosés or assorted sparkling wines make for interesting and versatile complements to all kinds of dishes, with a full-bodied Chardonnay being the exceptional white wine choice, capable of handling heavier meat options when a red wine has been vetoed.

By picking wines with the best food-pairing potential in mind, you can maximize the wine's ability to support the food ordered.

The best restaurant wine values tend to be at the midrange price points. A restaurant's cheapest wines are often the most marked up. You'll recognize these right away—they're the wines offered by the glass at the same price you could buy the whole bottle retail. So for a better value, stick with wines that are moderately priced for maximum gain, and steer clear of the very cheapest restaurant wine options.

The Restaurant Wine Ritual

Each restaurant takes a different approach to the wine ordering and presentation ritual, yet you can expect several consistent themes when ordering a bottle at wine-savvy operations.

The Order

Wine lists have evolved from leather-bound tomes with fancy handwritten wine names, to easily updated lists created with laser printers, to electronic tablet menus.

Ordering wine in a restaurant shouldn't be much different from picking up a bottle at your local wine shop. Sure, there can be the "pressure" of picking the right wine for the group or to pair with a dish, but it's still about personal wine preferences. Pick the starters or entrées first and then match the wine. Keep basic pairing principles in mind—lighter wines with lighter fare and sauces, bold wines with heavier fare, sweet wines with spicy dishes, and bubbles with just about anything—and you'll be just fine.

If indecision still lingers, ask for help. Most restaurants equip serving staff with tried-and-true pairing combinations for key menu items. Some upscale restaurants also have a sommelier available to guide you through the wine list.

The Presentation

After you've ordered and the wine bottle is brought to your table, the server should take a brief moment to show you the bottle label. This is a quick check so you both can ensure the right wine is being opened.

The Cork

After the server opens the bottle, he or she might give you the cork. This gesture is a remnant of restaurant tradition, when the cork was given to customers to sniff to determine if the wine was bad and double check that the wine in the bottle was in fact the wine stamped on the cork. Decades ago, unscrupulous proprietors might refill a pricier wine bottle, typically from the higher-end French Bordeaux or Burgundy houses, with a lesser wine and pop in a new cork—one that was missing the producer's or estate's name—pulling a fast one on unsuspecting patrons.

What should you do with the cork? Do you sniff it? No. Do you pick it up and inspect it? That's up to you. The best bet is to leave it and move on. If something is off with the wine, you'll detect it in the glass easier than by smelling the cork. However, if you must check the cork for something, you can verify the producer's name is stamped on it and that one end is stained and still moist from the wine, indicating the bottle has been stored on its side at least for some portion of its life.

The Taste

Now for a quick taste test. Your server will pour just a splash of wine in your glass. Give it a good swirl, take a generous sniff, and taste it. If the wine is great, say so. If not, ask a question or two.

The most common wine faults are oxidation, corked, reduction, or refermentation. Corked wines smell musty, like wet newspaper or cardboard that's been left out a while. Wines that have been improperly stored or have oxidized will have off colors, taste flat, and lack flavor. Reduction (a lack of oxygen) can lead to sulfur smells, and refermentation (when an unintentional second fermentation takes place in the bottle) shows up with a slight carbonation and cloudy color to the wine.

The Nod

If you approve of the wine, say so and the server will fill the glasses.

If you think there's truly something wrong with the wine, say so—*now*—and see what can be done to correct the situation. Most restaurants want you to enjoy your wine and will work to get it just right. But you can't sip through half the bottle before you decide the wine is flawed.

If you decide the wine isn't something you particularly like, but it's not technically flawed, just chalk it up experience and make a mental note to either check out a similar wine by another producer or from another region or steer clear of the wine altogether in the future.

More and more restaurants are presenting their wine lists on iPads or other tablets and letting customers select a wine from the screen. This also gives customers the opportunity to do their own research on the wine's origins, what kind of ratings it has received, the winemaker's notes, and pairing tips—all via touch screen.

SERVING WINE

Drinking wine is easy, but serving it can take a bit of effort. Choosing a bottle, opening the bottle, deciding whether to decant the wine or not, and catching the wine at the right serving temperature—there's plenty to consider. And that's all before you even look at glassware protocol or take the first sip. Wine is a high-maintenance beverage, but it's worth it.

A mind-boggling array of wine gadgets await eager enthusiasts: fancy bottle openers, electronic wine chillers, wine glass thermometers, traditional (and peculiar) decanters, pour-through aerators, refrigerated wine storage units, and even full-blown wine cellar cabinet systems. However, what you actually need to serve, taste, and store wine is pretty straightforward. A corkscrew, some wine glasses, a decanter (any jug will do), and a cool dark place to keep your wine are enough to get you started. Keep serving temperatures and appropriate glassware in mind when pouring, and you've just raised your wine experience another notch.

CORKSCREW PROTOCOL

Today's wine bottles essentially have three closure options:

- Natural corks
- Synthetic corks
- Screw caps

Traditional cork made predominately from Portugal's cork oak trees still thrills with classic culture and custom, but synthetic corks and screw caps are gaining ground as easy solutions to cork taint (a chemical contamination associated with natural cork closures that makes wine smell musty and moldy).

Opening a Bottle of Wine

Opening a bottle of wine with a screw cap closure is pretty straightforward and simple. Just twist it off.

Opening bottles with corks, whether natural or synthetic, isn't difficult. Cork-popping tools abound, but the ubiquitous "waiter's friend" style of corkscrew is the most familiar, reliable, and compact. The waiter's friend resembles a foldable pocketknife with three moving parts: the mini capsule-cutting blade; the actual 2-inch (5cm), 5-spiraled screw (or "worm" as it's also called); and the metal lever.

Here's how to open a bottle of wine with a corkscrew:

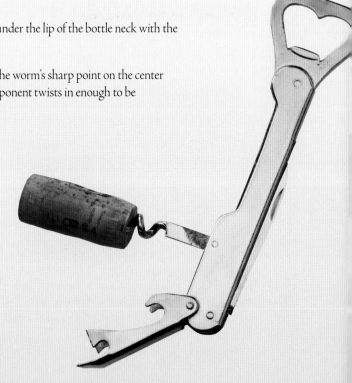

1. Cut the wine's plastic or foil capsule, or cover, right under the lip of the bottle neck with the corkscrew's mini blade.

2. Fold the blade in and then unfold the worm. Place the worm's sharp point on the center of the cork, rotating the handle until the worm component twists in enough to be completely covered by the cork.

3. Open the metal lever and secure the notched footing to the top edge of the bottle's upper lip, keeping it steady and in place with your other hand.

4. Slowly and gently pull up with the handle, carefully rocking the cork from side to side as you keep a constant upward motion until the cork clears the bottle.

Remove the cork from the worm, and you're ready to pour.

Cork Contingencies

Broken corks happen. So when it happens to you, just re-route the worm into the cork at a slightly off-center and diagonal angle, making every effort to run the worm clean through to the bottom of the cork. Then extract as you normally would with a touch more wiggle action if the cork seems impacted against the sides of the bottle neck.

Corks can also crumble or get pushed all the way into the bottle. You might be able to easily remove the cork crumbles by fishing them out of the poured wine with a spoon. If the cork has disintegrated into sawdust instead of sizeable crumbs, pour the wine into a decanter or carafe while straining through a mesh tea strainer or coffee filter.

Corks that have been pushed clear into the bottle can be tricky. Pour the wine into a decanter to serve if you don't want to deal with the bobbing cork, or just pour the wine with the cork still in the bottle. Hold the cork clear of the bottle neck when you pour by using something thin like a shish kabob skewer.

The Cork Is Out!

Now that the cork is out, it's time to pour. The typical 750-milliliter wine bottle holds just over 25 ounces. A standard pour is 5 ounces (148 milliliters), so there's just enough wine in the average bottle for five separate servings.

However, before you pour the wine, you have several factors to consider: serving temperatures, serving order, and whether you need to decant the wine or not.

Wine bottle closures are a hot topic for debate these days, with traditionalists preferring natural cork closures, especially for high-end wines intended for long-term aging. On the other hand, many white wine producers, especially in Australia and New Zealand, favor the screw cap for maintaining the wine's lively, fresh appeal. Admittedly, screw caps have had to overcome a "cheap" wine stereotype, and many early versions certainly supported this theme, but in today's market, countless mid-range and high-dollar exceptions would rather risk perception issues than cork taint in their wines.

SERVING TEMPERATURES

Too often, white wines are served too cold and red wines are served too warm. When we automatically serve white wines "well chilled" and red wines "at room temperature," we can miss out on subtle—and not so subtle—improvements that come when serving red wines cool to the touch and white wines without the deep freeze.

Red wines served too warm can come across fairly harsh, showing tight tannins and excessive alcohol on the palate. White wines served too cold can become muted both in their aromatics as well as their flavor components. That said, white wines should be served cold, but if you find it's too cold to allow the wine to open up properly, let it sit a bit in the glass to warm up.

You can easily chill white wines, as well as rosés or sparkling wines, by the ice bucket method. Simply fill a decent-size bucket or an empty jug half full with ice, place the wine bottle in the bucket, and fill the bucket to the top with cold water, making sure to cover the wine bottle neck. Leave the wine in the bucket for 20 minutes, and voilà—a perfectly chilled bottle of white wine.

Here's a quick cheat sheet of serving temperatures for different types of wines:

Champagnes and sparkling wines: 45°F (7°C)

White wines: 45°F to 60°F (7°C to 15.5°C; less expensive, lighter styles need lower temps)

Rosés: 50°F (10°C)

Dessert wines: 60°F (7°C)

Red wines: 60°F to 65°F (7°C to 18°C)

Fortified wines: 60°F to 65°F (7°C to 18°C)

If the red wine bottle doesn't feel cool to the touch, stick it in a bucket of ice for a few minutes or the refrigerator for 10 minutes. Remember that the lighter-bodied, low-tannin reds like Beaujolais, Pinot Noir, and some Spanish Riojas can take more chill with exceptional results.

If you plan on enjoying seconds of the wine, keep white wines in the ice bucket between refills. Reds might need a quick refresher prior to the pour if the bottle feels warm to the touch.

SERVING ORDER

If you're going to be enjoying several wines in a single sitting, pay attention to serving order. It's best to serve and drink white wines before red wines because the tannins and heavier overall structure of reds can make the palate transition back to the relatively lighter styles of whites tricky.

Aim for a general sequence of light-bodied to full-bodied wines to make discerning the delicate nuances of the light- to medium-bodied wines easier at the beginning of the tasting, before you bring on the palate power of a heavier wine.

With white wines, maybe start with a Riesling (light-bodied) first, then a Sauvignon Blanc (medium-bodied), and finish with a Chardonnay (full-bodied).

Red wines can be trickier to determine body style, with different reds presenting themselves in medium-bodied or full-bodied profiles, depending on various grape and wine-making factors. In very general, red wines with a lighter body include Beaujolais, Pinot Noir, Chianti, Rioja Crianza, and less-expensive Burgundy. Medium-bodied reds include Cru Beaujolais, Côtes du Rhône, Crozes-Hermitage, Chianti Classico, better Burgundy, some Merlot, some Cabernet Sauvignon, some Zinfandel, some Syrah, Rioja Reserva, and some Malbec. Fuller-bodied reds include Italy's big boys Barolo and Barbaresco, high-end Bordeaux and Rhône Valley reds, Priorat, Douro reds, Petite Sirah, and Merlot. Cabernet Sauvignon, Zinfandel, Syrah, and Malbec may also weigh in on the fuller-body scale. If in doubt, check the alcohol levels, and order the wines in ascending order of alcohol.

If dessert wines are in the tasting mix, save them for last, after you've enjoyed the dry wine styles.

To Decant or Not to Decant?

Decanting, or pouring wine into a special glass decanter or simple carafe, serves two primary purposes. It increases the surface area of the wine, allowing more aeration, which helps tame tight tannins and opens aromatics and flavor profiles in young or highly tannic red wines. If you're drinking a youthful red wine like some California Cabs or Merlot, Bordeaux blends, Rhône Valley reds, or more mature reds with bold structure like Italy's big Barbarescos, Barolos, and Super Tuscans or Spain's Priorat, decant the wine for an hour with the younger wines and several hours with the highly tannic Italian or Spanish reds to maximize aeration.

Decanting also makes removing sediment—the nasty, gritty solids that can accumulate at the bottom of a bottle after many years of aging—easier. By pouring the wine into a decanter, you can watch for the sediment to kick up and cloud the last few ounces of wine in the bottle. With a bright light illuminating the bottle, it's easy to see when the sediment cloud reaches the bottle neck so you can stop pouring.

Older bottles—we're talking 10+ years—need to stand upright a day or two before the big pour to allow the sediment to settle to the bottom of the bottle. Also, when decanting an older wine, it's the sediment you're trying to remove, so they don't need a lot of extended time in the decanter. Well-aged wines become more fragile over time, and the extra dose of oxygen can start rapid deterioration. Younger red wines with heavier profiles benefit from a bit of aeration overall to tone down the tannins and open the aromas.

For quick aeration, especially for a young red, consider a mechanical aerator. With this simple gadget, you pour the wine through the aerator into the glass. Popular models include the Vinturi Aerator, Metrokane Rabbit Aerating Pourer, and the Soiree Bottle-top Aerator.

GLASSWARE

Today's wine glass options offer tremendous variety, from trendy stemless glassware to the traditional full-stem, medium-bowled wine glasses that have been the industry's hallmark for decades. A good wine glass allows a clear view of the wine's color components, provides for easy swirling to encourage the wine's innate aromas to open, and is a pleasure to drink from thanks to a thin, tapered rim.

Rim

A wine glass's rim should be thin and well tapered. An elegant rim that's thinner than the glass itself makes for a glass that's easier to drink from, and the tapered edges concentrate the wine's intrinsic aromatics.

Bowl

The ideal bowl is smooth and crystal clear and offers room to accommodate a full serving of wine with sufficient space to swirl without spilling. The bowl's size varies considerably depending on the style of wine it's designed to hold. Red wine glasses are the largest, and Port wine glasses are the smallest. Red wines need the larger bowl to maximize air contact and surface area, whereas Ports are generally sipped in smaller quantities and tend to be deeply concentrated.

Stem

The stem allows for easy holding, without constant contact with the bowl.

Base

The base provides a firm foundation for the wine glass. Some advocate holding the glass by the base to keep your hands as far from the bowl as possible. Touching the bowl can smudge the glass, impede your view of the wine's color nuances, and even warm the wine.

White wine

Size:
7 or 8 inches
(17.75 to 20.25cm)

Capacity:
8 to 12 ounces
(236.5 to 355ml)

White wine glasses are typically tapered at the top in an effort to concentrate the delicate aromas of the wine. The smaller bowl is intended to keep the wine chilled and help deliver the wine to the center of your palate. Exceptions would be the larger bowls of Burgundy glassware designed to allow sufficient room for the hallmark aromas to develop, accent the wine's complexity, feature its full body, and accommodate the higher alcohol levels.

Red wine

Size:
7 to 10 inches
(17.75 to 25.5cm)

Capacity:
8 to 20 ounces
(236.5 to 591.5ml)

Red wine glassware is characterized by a large, round bowl. The design amplifies the aromas and expands the flavors of the wine by allowing the wine to breathe, effectively increasing the wine's surface area to encourage the mingling of the wine with oxygen.

Champagne

Size:
8 to 10 inches
(20.25 to 25.5cm)

Capacity:
6 to 10 ounces
(177.5 to 295.75ml)

The Champagne glass is noticeably elongated with a tall, narrow design that highlights and preserves the wine's bubble structure while also serving to capture the engaging aromas of a sparkling wine.

Stemless

Size:
3 or 4 inches
(7.5 to 10cm)

Capacity:
8 to 20 ounces
(236.5 to 591.5ml)

Stemless is the latest in trendy glassware. With its contemporary, no-frills design, stemless glassware is easier to store, wash in the dishwasher, and keep from tipping over.

Port

Size:
4 or 5 inches
(10 to 12.75cm)

Capacity:
4 ounces
(118.25ml)

Fortified wines like Port are higher in alcohol and offer dense, spicy aromas. Glasses with smaller capacities are preferred for Port and other fortified favorites, as the serving sizes (3.5 ounces/103.5 milliliters) tend to be less than table wines and the aromas are already quite concentrated.

WINE TASTING ESSENTIALS

The difference between *drinking* wine and *tasting* wine can be dramatic. When simply drinking wine, you tend to take in the total package—how it tastes overall, not necessarily what you taste specifically, or how the wine pairs with food in general, not distinguishing components that marry well with particular flavors.

When tasting wine, on the other hand, you take a moment to engage with the wine on several sensory levels. You check out color characteristics and stop to smell the roses—and the fruit, spice, veggies, nuts, earth, smoke, herbal, or toasted notes. Then you sip, slowly, waiting for textures, flavors, and the weight of the wine to make their own unique palate impressions.

To truly taste a wine, you must take the time to look, sniff, sip, and then evaluate.

THE LOOK

Take a good look at your wine. What color is it? Color can communicate quite a bit about a wine, including its age, hints on grape types, and the general condition, along with clues about the wine's body and weight structure. It seems easy enough to just tilt your glass over a white background, be it a napkin, a tablecloth, or paper, and check out the wine's color. But moving past simply "red" or "white," "rosé" or "sparkling," "fortified" or not, you need to start asking some questions:

- What shade of red? Is it purple, ruby, garnet, or brick red in color?

- What color of white? Clear, straw, golden, or amber?

- Can you see through the wine? Just how translucent is it? If not opaque, how is the color density, especially when compared to another wine?

It's the grape's skin that determines a wine's final color scheme. Squeeze a black grape and a green grape, and the juice runs clear from both. However, if the juice from a black grape spends a good bit of time soaking in its crushed grape skins, the color pigments in the skin "stain" the juice. In the wine world, this process is called *maceration*.

The thicker the grape's skin, the more color pigment it can contribute. Just as the longer the juice is in contact with the skin, the more color can bleed into the juice. That is why thin-skinned grapes like Pinot Noir typically make more translucent wines, while thick-skinned grapes like Syrah bring more concentrated color to the glass.

Wines can also change color as they age and oxidize, with reds exchanging their bright burgundy shades for more mellow orange to brick tones. Whites gain darker, burnished gold and brown hues with more time in the bottle. If you've got a fairly young wine (a few years or less from the vintage date) with "old wine" colors kicking, there's a pretty good chance the wine has been exposed to considerable oxidation via a hairline crack in the cork or lousy storage conditions.

A wine's color also conveys expectations on whether the wine is light, medium, or full bodied in nature. The lighter, translucent end of the color spectrum hints at light-bodied wines, midrange color tones predict medium-bodied sips, and deeply concentrated colors in both the red and white range can indicate full-bodied beasts.

THE SNIFF

Your sense of smell is your greatest asset when it comes to tasting wine. That's right—your sense of *smell*. Your taste buds are fairly limited in their sensory uptake, detecting sweet, salty, sour, and bitter, but your nose picks up thousands of unique scents and partners with your tongue to convert those scents into full-on flavors.

This is where the professional wine swirl comes in to play. (Double check that you have enough room to swirl the wine in the glass without spilling it. This may be the only time a glass that's half full can be perceived as a *good* thing!) By seriously swirling a glass of wine, you aerate the wine, allowing oxygen to interact with all those little aromatic molecules captured within the glass. As the wine's alcohol begins to vaporize, the aromas come with it. Give the glass a good 10-second swirl to lift the scents up and out of the glass and then stick your nose way down *into* the glass and take a couple quick sniffs.

What do you smell? Sniff again. Is it somewhat familiar? If you're examining a white wine, do you smell white fruit? If you're sniffing a red wine, do you pick up dark or red fruit scents? Can you detect specific fruits, flowers, and spices, or do the aromas veer toward more general impressions of sweet, floral, herbal, woody, earthy, or outdoorsy smells?

If you aren't able to pick up much in the scent department, double check your swirl power. You'd be amazed at the side-by-side aromatic differences between a wine that's been well swirled and one that hasn't. Don't skimp on this critical flavor-releasing factor.

How does wine made solely from grapes smell like flowers, fruit, spice, or other more exotic, nongrape scents? A number of fermentation factors play into a wine's ultimate aromatic appeal.

When yeast metabolizes the grape's sugar into alcohol, thousands of chemical compounds—mainly aldehydes and ketones—are also being formed from the grape's natural flavors and structures. These compounds create the aromatic metabolites that debut as the familiar fruit, floral, and vegetal scents often associated with classic wine aromas beyond the scope and smell of a simple grape. You're likely to detect plum, cherry, strawberry, blackberry, and raspberry, among other red and black fruit scents and flavors found in red wines, and citrus, apple, pear, mango, pineapple, and other "white" fruits in the aromatic scope of white wines. These are all aromatic metabolites of the starting sugars that happen to echo and share similar chemical structures, smells, and flavors of the fruit or floral scents they reflect. Some people ask if additional fruit juice, fruit essence, or fruit itself has been added to a wine to promote a distinct fruit-themed scent and taste. The answer is no. It's just fermented grape juice in all its aromatic glory.

Oak is another primary player in the aroma arena. Used to season the wine with smoky nuances, baking spices (vanilla is often the most recognizable), or rich buttery tones (especially in Chardonnay), oak's mighty influence manifests itself in many discernible forms. By amplifying a wine's aromas, adding significant depth, complexity, and underlying richness, oak strives to complement the charismatic chemistry of fermentation.

A wine's aromas may also be broken down into primary, secondary, and tertiary aromas. The primary aromas are formed by the grape itself, a direct result of the individual grape type or types used to make the wine. Secondary aromas encompass the new spectrum of scents created during the fermentation process like various fruit and floral notes. Tertiary aromas form after fermentation as a result of oak or bottle aging. These might run the lines of oak-induced vanilla or clove spice. Technically, the secondary and tertiary aromas found in a wine may also be referred to as the wine's "bouquet."

The numerous scents buried in a glass of wine are collectively called the wine's *nose*.

The Sip

The sip is where it all comes together. Although your nose is involved in flavor detection, your tongue deserves some credit, too. Your tongue helps you pick up four taste sensations—sweet, sour, salty, and bitter. The tip of your tongue tastes sweet (or dry), the sides pick up sour (tart acidity in wine), the middle takes salty (not in wine but in the foods that accompany it), and the back brings bitter flavors to taste (think tannins).

To see how this works, begin by taking a decent sip of a wine and hold it in your mouth a moment or two before swallowing. Allow the wine to roll around, hitting all areas of your tongue, and feel the intrinsic textures.

Is the wine dry (which is the opposite of sweet in wine words)? How dry? Bone-dry, semi-dry, or turning sweeter? Are the tannins silky, velvety, harsh, or overly drying in the red wine? What about the acidity (found in both red and white wines although it tends to burn brighter in whites)? Does it provide structure for the fruit, or is it overly aggressive on the palate? Does it make your mouth water? Is the wine's weight light or heavy on the palate? How's the fruit? Forward and more pronounced, somewhat subtle and understated, or lacking altogether?

A wine's components—sugar, tannin, acidity, and alcohol—should all be in complete harmony. If one piece stands out and demands all your sensory attention, the wine lacks balance. If the wine runs "hot" and all you smell and taste is alcohol, even after chilling it a bit, you've got a wine that's off balance. The red wine's tannin content, which comes from the grape's skin, seeds, and stems, can be tough to keep in check and is one reason why particularly tannic wines are generally better with some age to relax their high-strung structure.

In the end, a wine's innate flavors are derived from the ground and grape itself, the vigorous process of fermentation, the influence of oak, and the age of the wine. Older wines soften around their tannic edges and lose some of their pronounced fruit character, yielding to the development of tertiary aromas and greater flavor complexity.

Technically, taste is broken down into three steps:

The attack: This is the very first impression the wine makes when you take a sip. The four components of wine—sugar, alcohol, tannin, and acidity—work out initial sensations on the palate. They don't communicate true flavors (like fruity or spicy), but instead highlight textures, concentration, and intensity. Tannin makes your mouth feel dry; acidity makes your mouth water. Is the wine sweet or dry, light-bodied or full-bodied, crisp or creamy, simple or complicated? Typically, the more alcohol a wine has, the fuller the body will feel.

The *body* of a wine refers to its innate weight in your mouth. Light-bodied wines are a similar weight to skim milk, medium-bodied wines stick closer to whole milk in mouthfeel, and full-bodied wines can seem as heavy as fresh cream to the palate.

The evolution: This is where flavors form. The evolution phase, also called the "mid-palate," is where the specific fruit flavors evolve in the mouth as scents and tastes collide. If it's a red wine, do you detect ripe berries, plums, or cherries? Is it earthy, fresh, or spicy? In a white wine, how's the fruit? Does it greet with plenty of citrus, stone fruit (apricot, peach), tree fruit (apple, pear), or tropical delights? What about butter, honey, and herbs? Can you taste them?

The finish: In this final phase, think about how the wine finishes. Do the impressions fade quickly away after you swallow (a short finish) or do they linger longer (a long finish)? What was the final flavor? Do you want to take another sip or just let it sit?

So did you like the wine? Would you buy it again? What kind of food would delight with this wine?

You can do the sip test with a single wine or with multiple. Wine tastings, where different wines or vintages are simultaneously compared and contrasted with one another, can be instrumental in gaining a real feel for a wine's unique components and offer a practical palate primer on different grape varieties.

Vertical tastings take the same wine from various vintages and line them up side by side so you can see how the wines express themselves uniquely based on the growing year. A horizontal tasting takes several wines made from the same grape and same vintage but from different producers to compare wine-making influences. Varietal tastings showcase different grapes, generally all whites or all reds, and evaluate the distinct aromas, flavors, and structures found within the lineup.

Factors for Pairing Food and Wine

Wine is made for food. It's a noble backdrop, a spice or seasoning, for food. It strives to bring out the best in the food it's well partnered with, and it protests pairings that bring a harsh, bitter voice to both. Acidity is the key to food's deep affinity for wine, with the ability to heighten flavors, perk up bland dishes, and add interest to otherwise unremarkable ingredients. And that's not to mention the foundational physiology found in sipping a wine with decent acidity and the mouthwatering phenomenon that renders a palate ready to receive the next bite.

The delicious, dynamic relationship wine pursues with food cannot be overemphasized—they truly go hand in hand. Wine and food are made for one another, and just like a good marriage, both are bettered by the partnership. In the best food and wine pairings, both the dish and the wine are elevated and noticeably improved in terms of flavor, satisfaction, and component interaction. Simply stated, they both taste better together than they do solo.

Poor pairings often turn bitter, with astringent qualities surfacing, tannin dominating, or acidity raging a sharp and cutting edge on the palate. Match dry, crisp wines with decent acidity to something sweet, and they revolt with sharp tartness. Or taste a tannic red in the presence of spicy fare, and witness the harsh, astringent, mouth-drying tannin and dominating alcohol that result. Thankfully, plenty of food and wine pairing combinations work extraordinarily well together, and with just a few guidelines, you'll be well on your way to pairing paradise.

Food should taste better and wine should taste smoother when they're paired well together. Bitter, sour, and overly tart notes are red flags the relationship isn't working. If that happens, just pick another wine or take a different bite and try again. Good pairings exhibit a synergy that renders each partner better in the presence of the other, like peanut butter and jelly, strawberries and cream, or ice cream with chocolate sauce.

PAIRING PRINCIPLES

The weights, flavors, textures, and individual components of food and wine bring about an ever-changing pairing chemistry that promises to either complement, contrast, or clash. Finding common ground with both the food and the wine sets the initial pairing stage.

Weighty Matters

Matching the perceived weight of the food with the weight of the wine ensures neither the food nor the wine dominates the aromas, flavors, or textures of the other. This is where the most prevalent (and popular) piece of food and wine pairing wisdom takes firm root—you're probably already familiar with it: "white wine with white meat" and "red wine with red meat."

Although this is a terrific starting point, it falls woefully short in the long term, especially for adventurous foodies clamoring to ignite a food-and-wine-pairing frenzy with the latest ethnic inspirations or the headliner grape from an up-and-coming wine region. Trust your palate. Taste is an individual endeavor, and what works for you might not work for the local wine expert, and that's just fine.

Tactile Textures

Weight isn't the only consideration; think about textures, too. Wines with soft, silky textures tend to be delightful companions with food that shares a similar tactile structure. Rich, fatty foie gras and the sweet, plush textures of a full-bodied Sauternes makes a classic pairing for a reason. The weight *and* the texture are the same.

Likewise, the light to medium body of fresh shellfish or grilled fish requires kindred weights in wine like Pinot Grigio, Sauvignon Blanc, Grüner Veltiner, or Bordeaux Blanc.

From red wines such as the softer profiles of Pinot Noir, perfect for salmon or roasted rack of lamb, to the big bodies of California Cabernet Sauvignon begging for the best cuts of beef, the palate weight of the wine should be in the same league as the general weight found on your plate.

Favored Flavors

Flavors may be the most obvious factor in establishing the tastiest food and wine pairings, but they can be viewed from two distinct angles: flavors that complement and flavors that contrast.

Complementary flavors run the "like with like" themes. Joining the crisp, herbal, and zesty acidity of Sauvignon Blanc with the fresh, pungent aromas and tanginess of goat cheese showcases how the tart and tangy flavors are subdued to bring about the bright, fresh flavors in both the cheese and the wine without doubling up on the palate "zing." Lobster drenched in butter matches up wonderfully with a big, buttery Chardonnay—like with like.

Contrasting flavors work a bit different. For example, what happens when you partner a sweeter-styled wine, like an off-dry Riesling, with some serious spice in feisty Asian fare? The sweet and spice don't wage war in your mouth. Instead, they manage to find a delicious middle ground, with the sweet subduing the spicy flames and the spice toning down overambitious sugar.

Many cultures place a priority on their unique sauces that accompany their cuisine, and it's absolutely critical that the wine pairing does the same. If you're pairing a wine with a dish that comes covered in a "special" sauce, try to break down the dominant components of the sauce for pairing consideration.

CLASSIC PAIRINGS

Cultures have created cuisines based on their current, seasonal sources of food throughout history. Happily, regional wines have also been crafted and honed to support and accent the best of the local fare, with some classic pairings standing the test of time because they're delicious from start to finish. These classic pairings may spark experimentation or inspire new creations, but they remain the bedrock of food and wine pairing protocol.

Red Wines with Steak

The heavy tannin content in various red wines marries particularly well with the high protein and fat found in steak. The tannin structure combines with the fat and protein to become less concentrated, rendering the tannin more mellow and manageable.

Red Bordeaux with Lamb

The earthy, terroir-driven nature of Bordeaux's red wines can display a gamelike quality, with subtle fruit and deep-seated tannins. When combined with the intense, earthy, and similar gamey flavors of roasted lamb, the wine brings a tremendous finesse and complement to the dish.

Pinot Noir with Duck

Pinot Noir handles all sorts of poultry picks with classic finesse and silky tannins, but the richer flavors of duck meat spotlight Pinot Noir's cherry, raspberry, and often earth-driven profile with ease and elegance.

Sauternes with Foie Gras

This is a perfect example of rich meets rich, where the ultra decadent and completely complementary textures of both the food and the wine line up to bring a savvy palate synergy, while also showcasing the dynamics of matching a sweet wine with more savory fare.

Chablis and Oysters

Chablis is an elegant, bone-dry white wine made from Chardonnay grapes in Burgundy with delicious mineral-laden acidity that's rarely oaked. Like a fresh squeeze of lemon, the subtle citrus fruit, lean flinty character, and palate-cleansing acidity of Chablis makes a magical marriage with the succulent textures and salty, briny nature of fresh oysters on the half shell.

Champagne and Caviar

Light- to medium-bodied Champagne, with its crisp, vibrant acidity and bright bubbles, cuts right through the oil and salty flavors of traditional caviar, improving oily textures and augmenting briny flavors along the way.

Sauvignon Blanc and Goat Cheese

A stunning, super-savory marriage of food and wine, Sauvignon Blanc (Sancerre and Pouilly-Fumé in France) brings out the very best in fresh, pungent goat cheese. The remarkable crisp acidity in the wine pulls out the fresh factor in the cheese and manages to subdue the tart, tangy nature in both.

Chardonnay with Brie

The creamy textures in both the cheese and the wine make for a natural pairing alliance where rich flavors and smooth, buttery textures rule.

Port with Stilton Blue Cheese

This very traditional pairing offers a crash course in contrasting components. The sweet-natured, full-bodied flavors of Port provide a stunning contrast to the salty, pungent aromas and equally full flavors of Stilton blue cheese. Rocking back and forth between sweet and savory with each sip and nibble, the pairing is undeniably powerful.

Cabernet Sauvignon with Parmesan Reggiano

The intensity of the tannins in the wine is toned down by the protein and fat content of the full-flavored cheese. The fat buffers the tannins, reducing the innate astringency and brings the wine and cheese together into a smooth semblance of fruit-filled flavor.

Cabernet Sauvignon is also an ideal pair with aged Gouda.

Vintage Port with Dark Chocolate

In a brilliant display of calming complements, the tannins in the Port and the energetic tannins in the dark chocolate call a truce, allowing the rich, concentrated fruit flavors of the wine to shine.

STORING WINE

Some people store wine for years, while others only keep it for a few hours. Either way, having a basic knowledge of how to house wine, whether for a decade or a day, saves you from sipping sour grapes.

Wine storage is a fairly straightforward endeavor, although today numerous options are available, ranging from a box in the basement, to simple wine refrigerators and larger temperature- and humidity-controlled cabinets, all the way to full-blown segments of a home solely dedicated to cellaring a pretty decent wine stash.

Whichever storage option you pursue, it should be a cool, dark environment where the bottles can rest on their side.

SMART WINE STORAGE

There's a reason why wines have been traditionally stored in underground cellars and wine caves. The optimal wine storage demands are easily met there: cool temps, lack of light, some humidity, and completely still surroundings.

Cool Temperatures

If you remember nothing else, remember 55°F (13°C) as the average optimal storing temperature for wine. Sure, a little wiggle room exists on both sides of that magic number, but if hard-pressed to set a dial on your wine cooler, this is it.

Excessive heat will damage a wine given a bit of time, and big swings in temperature are just as detrimental because they can cause the cork to contract, letting oxygen begin its devastating invasion.

No Light

Ultraviolet light can cause wines to age prematurely and spoil, rendering white wines out of sorts faster than reds. Light certainly isn't a friend to either, ever.

Humidity

Some moisture—in the 60 to 70 percent range—is great for long-term wine storage. It helps keep natural corks from drying out. In short-term situations (3 to 6 months), humidity is not something to be overly obsessed about.

Still Surroundings

Excessive vibration, like that found in or on a refrigerator, can shake up sediment and alter the chemical aging process of a wine. It's best to store wine bottles in a still, static environment and always on their side. The constant contact with the wine keeps the natural cork from drying out.

Short-Term Storage

Once opened, a bottle of wine begins to oxidize. Oxygen can be a good thing to aerate an uptight, tannic red wine, but it can be a wine killer given more time. Oxidation is ultimately responsible for a wine's deterioration, taking a wine that was fresh and fruity on day 1 to flat and funky on day 2, 3, or 4. Wines opened for a day or more start to lose their aromatic and flavor pizzazz, bringing muted fruit tones where there was once bright, ripe berry, and worn-out woodlike aromas where only the day before vanilla and spice emerged. However, several tricks can stall oxidation and keep an opened bottle humming for another day or two.

Refrigeration is mission critical to prolonging the limited life of the wine—for both red and white—so once the bottle is opened and the wine has been poured, stick the cork back in the bottle and refrigerate the wine right away (unless you're confident that the bottle will be completely consumed).

Several savvy wine-preservation gadgets promise to breathe a bit more lifespan into a bottle. Some function on a vacuum principle, like the Vacu Vin, which works to remove excess oxygen from the bottle prior to popping the cork back in. Others work on oxygen displacement and provide an inert layer of gas via a spray system to blanket the exposed wine surface and keep excess oxygen from reaching the wine. Both systems can make a difference in prolonging a bottle of wine when partnered with refrigeration.

Long-Term Storage

Conservatively, less than 10 percent of wines on the market improve with additional aging. Bordeaux, Burgundy, California Cabernet Sauvignon, Italy's Barbarescos and Barolos, some Australian Shiraz, and Spain's Priorat are the most common age-worthy wines. The majority of the wines built to age are red wines with high concentrations of tannins or white wines with exceptional levels of acidity.

Long-term storage, with 5 to 10 years being on the low end, is where the specific demands of constant, cool temperatures, decent humidity, and a dark, still environment really come into play.

Don't keep a wine "stored" in a refrigerator for safe keeping, thinking it's cool and dark, so it must be perfect. The "cool" is actually quite a chill, often running in the 35°F to 38°F (1.5°C to 3.5°C) range and flattening flavors and aromas in the process. The motor constantly vibrating your vino doesn't help, either.

WINES OF EUROPE

Wine has played a critical role in European history and economics. Roman soldiers planted vineyards during the expansion of the Roman Empire, and monks maintained vines during the medieval period. Wars were fought and vineyards trampled and subsequently divided during Napoleon's revolution, and wines were bought and shipped to Great Britain for centuries from Portugal or France, depending on trending alliances and enemies. America's founding fathers were fond of French wines and Portugal's fortified Madeira, which were perfectly crafted for the extended shipping across the rough waters of the Atlantic.

Wine has been an ongoing reflection of Europe's culture, cuisine, and history, an emblem of the place and people who worked the vineyards, grew the grapes, and bottled the wines. It's no wonder Europe at large focuses on *where* the grapes are grown not *which* grapes are grown. The tug-of-war over place name verse grape name is easily won by the profound respect for the land that produced the wine. So when you see a wine from one of Europe's most notable growing regions, be it in France, or Italy, Spain, Germany, or beyond, know that although the label might seem complicated with unfamiliar geography and long words that could be either villages or vineyards, there's a sense of pride and purpose in where the grapes are grown and that's precisely why the location adorns the label, not the specific grape. Place names, not grape names, rule on European wine labels.

European wine law dictates label requirements based on specific regional boundaries, grape varieties grown, vineyard management practices, aging protocol, and much more. The same grape types grown in Europe may exhibit more subtle, almost earthy components compared to their New World counterparts due to the specific soil, topography, climate, and winemaker influences that affect the grape.

Tradition has long held sway over technology in Europe's wine-making history. Yet the last few decades have witnessed an explosion in modernization of the wine-making and lab equipment alongside vineyard management techniques. A dynamic synergy between tradition and technology has created a wave of enthusiasm and consistent quality in Europe's select wine-growing regions. From family-owned boutique wineries to full-fledged corporate affairs, from everyday table wines to the high-end auction-bound bottles, the wines from Europe do their best to express the people and the place that brought them to life.

WINES OF FRANCE

You don't have to venture far in the world of wine before you realize France is a big deal. Most major wine regions seek to imitate the classic wines of France, and most serious wine consumers endeavor to navigate the complexities of regions, labels, and place names, as well as the complicated wine classification system intended to both maintain quality and communicate grape expectations to adoring fans.

France has earned its stellar wine reputation, built on the ritzy reds of both Bordeaux and Burgundy and the exalted bubbles of Champagne. The distinguished wines of France set the gold standard for many of the world's most influential wines. Alternating with Italy on a vintage-by-vintage basis for the globe's top spot as the largest wine-producing country, France maintains a significant impact on the viticultural (grape-growing) and vinification (wine-making) practices of both Old World and New World wine regions. On a global scale, France is first in wine consumption per capita, although the actual annual consumption has decreased by almost 50 percent in the last 50 years. Today, the typical French man or woman sips close to 50 liters (around 13 gallons) of wine annually.

From modeling compelling wine quality-control measures to being the first to embrace and educate on the concept of terroir (the collective effect soil, sun, topography, water cycles, and climate have on a grape-growing vine), bolstered by centuries of wine-making history and extraordinary tradition, with great diversity in both wines and regions, French wines are living legends. Despite the occasional high-dollar bottle, the majority of French wines do not come with legendary price tags. The fame (and fortune) of French wines has certainly been built on the big names, but most of the country's wines fall into wallet-friendly categories.

In this chapter, we begin by exploring the most prominent wine regions of France, and I share tips for understanding the various classifications, reading French wine labels, and pairing French wines with local food and beyond. (I've saved the Champagne discussion for later in the book, when we look at the whole of sparkling wines.)

FRENCH WINE CLASSIFICATIONS

Knowing and understanding French wines is critical to enjoying the wide world of wine to its fullest, yet French wine classification systems and resulting label layouts can get confusing quickly for consumers raised on New World wine labels highlighting grape names over place names.

Remember, the place, not the grape, is the priority when naming and labeling French wines. Bordeaux is a place—a city, in fact—right off the Atlantic coast. Likewise, Champagne is a place—a region just east of Paris. Specific grapes grow really well in certain places, making a location like Bordeaux famous for growing Cabernet Sauvignon and Merlot and a region like Champagne ideal for Chardonnay and Pinot Noir—which also happen to be the top grape varieties used for making the best bubbles of Champagne.

Let's take a closer look at French wine classifications:

AOC (*Vins d'Appellation d'Origine Contrôlée*) wines are at the upper echelon of the French wine classification system. These wines must meet key criteria in terms of vineyard and wine-making practices, as well as requirements for minimum alcohol levels, and they must adhere to rules for using only the permitted grape varieties grown within the specific appellation boundaries.

AOC wines often have a specific place name on the label between the words *Appellation* and *Contrôlée*. For example, a wine carrying a label from grapes sourced from a specific subregion of the famous Loire Valley like Sancerre reads *Appellation Sancerre Contrôlée*. At other times, the names of the general region or a small village appear on the label, followed by the initials *AOC*.

VDQS (*Vins Délimités de Qualité Supérieure*) wines are loosely translated as "wines of superior quality" and represent one step down in label requirements from the AOC wines. Often the vineyards may harvest grapes at higher yields, reducing the quality and concentration compared to the AOCs. The initials *VDQS* or the full-blown phrase appears at the bottom of the label.

Vins de Pays are considered "country wines" and are defined by a specific place or, more often, a large region. The rules and regulations for obtaining a Vins de Pays designation are significantly looser than the AOC or VDQS standards. The Languedoc-Roussillon region puts out plenty of good, everyday Vins de Pays wines each year, wearing the label *Vins de Pays de Languedoc-Roussillon*, inferring they're country wines from the French region of Languedoc-Roussillon.

Vins de Table indicates the simplest of all French wine. This label designation is used for the ubiquitous, everyday table wines, as the name suggests. Essentially, the wine just has to be made in France with French grapes—no specific place, no specific grape, and not even a specified year. These are typically low-cost, low-frills wines perfect for washing back a quick meal or packing for a picnic.

Bordeaux's best wine estates are classified by quality. Both Left and Right Banks maintain different classification systems, but the most famous classification came in 1855, ranking 61 estates into 5 categories: First Growths (the best of the best), Second Growths, and on down the line to Fifth Growths. Bordeaux First Growths (classified as Bordeaux's Top Chateau in 1855) include Chateau Haut-Brion, Château Lafite-Rothschild, Château Latour, Château Margaux, and Château Mouton-Rothschild (moved from a Second Growth in 1973).

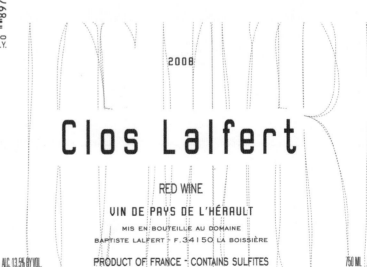

Parallèle **45**® CÔTES DU RHÔNE

—— APPELLATION CÔTES DU RHÔNE CONTRÔLÉE ——

This wine is named after the 45th Parallel, which crosses the Rhone Valley, passing through some of the estates and vineyards belonging to Maison Paul Jaboulet Aîné. **50% Grenache, 40% Cinsault** and **10% Syrah** give this wine its spicy, fruit-driven aromas. Best enjoyed young.

—————— www.jaboulet.com ——————

Mis en bouteille par **PAUL JABOULET AÎNÉ**

26600 LA ROCHE DE GLUN (France) - Produit de France - Product of France

ROSE WINE/ VIN ROSÉ

750 ml - 13 % Alc./vol.

GOVERNMENT WARNING : (1) ACCORDING TO THE SURGEON GENERAL, WOMEN SHOULD NOT DRINK ALCOHOLIC BEVERAGES DURING PREGNANCY BECAUSE OF THE RISK OF BIRTH DEFECTS. (2) CONSUMPTION OF ALCOHOLIC BEVERAGES IMPAIRS YOUR ABILITY TO DRIVE A CAR OR OPERATE MACHINERY, AND MAY CAUSE HEALTH PROBLEMS.

0CE10364

IMPORTED BY FREDERICK WILDMAN & SONS, LTD., NEW YORK, N.Y.

CONTAINS SULFITES

8 97441 75194 2

BORDEAUX, FRANCE

The Region and the Wine

Situated in the southwest corner of France along the Atlantic Ocean, Bordeaux boasts more than 10,000 wine producers and 300,000 acres (120,000 hectares), making it one of the largest and oldest French wine-growing regions. Geography is critical to understanding Bordeaux's wines. The region's waterways—primarily the Gironde Estuary, which forks into the Dordogne and Garrone Rivers—serves as the divider for the famous "Left Bank" and "Right Bank" wines, with the gravelly ground on the left (west) bank of the Gironde and Garrone good for growing Cabernet Sauvignon, and the heavier clay soils on the right (east) side of the Gironde and Dordogne perfectly poised for cultivating Merlot. "Right Bank Merlot" and "Left Bank Cabernet Sauvignon" are very general rules of thumb, and delicious exceptions exist on both sides of the bank.

■ Bordeaux region

Nearly 85 percent of Bordeaux's wines are red blends, and Merlot, Cabernet Sauvignon, and Cabernet Franc are the stars. Bordeaux's reputation is built on high-priced, age-worthy reds, yet most of the region's reds are reasonably priced and don't demand much cellar time. Wines grown in the Left Bank are typically blended with higher percentages of Cabernet Sauvignon for concentrated character and tighter tannins. Right Bank wines lean more heavily on Merlot in the blend with an emphasis on plush red fruit, smooth style, and silky tannins. Bordeaux wines aren't labeled *Left Bank* or *Right Bank,* so it's worth knowing key districts and towns on both banks. On the Left Bank, look for Médoc (district) and towns Listrac, Margaux, Pauillac, St. Julien, St. Estèphe, and Pessac-Léognan. On the Right, look for towns St. Émilion and Pomerol and AOC districts Fronsac and Côtes de Bordeaux.

Selecting and Pairing

Bordeaux's Merlot-based blends tend to be more approachable at a younger age with rich, red fruit; silky tannins; and a softer structure than their delicious, dark-fruited, tannic Cabernet Sauvignon cousins from the Left Bank. These innate wine features add to their food pairing versatility.

MEAT

From roasted leg of lamb to Bordeaux's best beef tartare, the red wine blends from both of Bordeaux's banks handle the heavy palate profile and protein content of red meat very well. For richer cuts of steak, try a Cab-based blend to allow the tannins to work their magic.

CHEESE

Cheddar, blue, and other aged cheeses with heavy aromas are a delight with younger Bordeaux reds as well as the heavier styles of Sauternes. Camembert, Brie, and goat's milk cheese marry well with the crisp acidity in many of Bordeaux's Sauvignon Blanc–based whites.

DESSERT

Sauternes is Bordeaux's famous dessert wine. From fruit-based confections, to almond pastries and pecan pie, to the locally made custard-centered Cannelés, the honey-filled flavors and full body of Sauternes turn up the volume in almost any dessert.

POULTRY

Roasted duck, goose, and other game birds are the perfect match for a Merlot-based Bordeaux red with softer tannins and ripe, red fruit nuances. Consider the region's rich foie gras for a perfect pairing with Bordeaux's decadent Sauternes.

SEAFOOD

Bordeaux's white wines (Bordeaux Blanc), made from Sauvignon Blanc and Semillon, make delicious food partners for the region's abundant seafood. Consider a dry, minerally white Bordeaux wine for pairing with salty oysters, scallops, shrimp, crab, and grilled fish.

PRODUCERS TO TRY

Chateau Coufran, Chateau d'Arche, Chateau Gazin, Chateau Greysac, Chateau Meyney, Chateau Olivier, Chateau Poujeaux, Chateau Paloumey, Chateau Rouget, Chateau Smith Haut Lafitte

BURGUNDY, FRANCE

The Region and the Wine

Tracing its way along the Saone River in east-central France, Burgundy's (*Bourgogne* in French and on the label) major growing regions cover just over 100 miles (161 kilometers) from top (Chablis) to bottom (Beaujolais), with close to 60,000 acres (24,281 hectares) under vine. Two grapes dominate Burgundy's wine stage, Chardonnay (a.k.a. white Burgundy) and Pinot Noir (red Burgundy). Burgundy sets the international Pinot Noir standard for taking the red grape to heavenly heights and commands global respect for its opulent Chardonnay. Keep a lookout for the place names of Chablis or Pouilly-Fuissé, two regions that produce world-class Chardonnay.

Burgundy region

Nowhere in the world does terroir shine as bright as it does in Burgundy, where two vineyards with just a road between them can cultivate grapes that offer dramatic differences in flavor, character, and composition. Burgundy's vineyards are notoriously small, fragmented plots of land often with numerous owners, a result of Napoleon's postrevolution dictates for the land and subsequent heirs. At just over 30 miles (48.25 kilometers) long, the growing region named *Cote d'Or*—literally "golden slope," for its famous fall foliage—is one of Burgundy's best wine-producing areas, with legendary white and red Burgundy coming from the villages of Meursault, Puligny-Montrachet, and Chassagne-Montrachet (search for these village names on Burgundy labels).

Selecting and Pairing

Pinot Noir may be the most food-friendly wine on the planet, offering tremendous pairing versatility, while Chardonnay is content being the pairing partner of choice for dishes with heavy butter and cream-based sauces as well as a faithful friend to seafood and poultry.

MEAT

Beef Burgundy (*Boeuf Bourguignon*) begs for a red from the region, so consider a Chassagne-Montrachet or Puligny-Montrachet. Grilled, stewed, and roasted beef or game, topped with mushrooms and red wine sauce, also marry well with the elegance and earthiness of Burgundy's reds.

CHEESE

The bright red fruit and subtle tannins of red Burgundy call for more mild cheese-pairing options. Consider Volnay with Brie, Camembert, Swiss, or Gruyère. Or take a solid-structured white blend from Pouilly-Fuissé and partner it with Gouda, Jack, and Brie.

DESSERT

Crémant de Bourgogne, a creamy, sparkling wine from Burgundy, offers the region's best bet for pairing with dessert. The bubbles and solid acidity complement rich cakes, strawberry-inspired spreads, local sweetbreads, and flakey butter pastries.

POULTRY

Roasted chicken and pasta with a cream-based sauce partner exceptionally well with a medium- to full-bodied white Burgundy, like Chablis or Pouilly-Fuissé, and the classic regional fare of *coq au vin* is a perfect pairing for one of Burgundy's red wonders from Pommard or Meursault.

SEAFOOD

Classic Chablis can run a little leaner, with less obvious body and more minerality than its New World Chardonnay counterparts, making it a top pick for pairing with fresh lobster, grilled fish, baked clams, sautéed shrimp, and smoked salmon.

PRODUCERS TO TRY

Bouchard, Christian Moreau, Faiveley, Fuissé, JJ Vincent, Joseph Drouhin, Louis Jadot, Louis Latour, Olivier Leflaive

RHÔNE VALLEY, FRANCE

The Region and the Wine

Located in the southeastern corner of France along the Rhône River, the Rhône Valley enjoys a sunny, Mediterranean climate. The warm weather conditions coax the grapes to fully ripen, producing abundant flavor, often higher alcohol, and natural complexity in the regional varietals. Most view the Rhône Valley as the Northern Rhône and the Southern Rhône, with the top two grapes being Syrah and Grenache. Powerful, full-bodied Syrah is the dominant red grape from the north and is also found in Châteauneuf-du-Pape, the south's most famous growing region. The prevailing grape of the Southern Rhône is the very versatile Grenache. It's typically blended with Syrah, Cinsault, and Mourvèdre.

■ Northern Rhône region
■ Southern Rhône region

Like much of France, the Rhône has its own wine terminology. The most well-known wine subregions (and label designates) from the Northern Rhône include Hermitage, Crozes-Hermitage, and Côte-Rôtie (also label designates), and the Southern Rhône is famous for the rich, fruit-forward reds from Châteauneuf-du-Pape (literally "the Pope's new castle"). The stunning color of Rhône rosés, dubbed Tavel, comes from soaking the Grenache skins with the juice for a short time. Essentially, Tavel brings the fresh fruitiness of a red wine in the style of a white with zesty acidity and easy pairing personality.

Selecting and Pairing

Red wines make up more than 90 percent of the Rhône Valley's wine scene; however, white wines, using Viognier (Condrieu), Picpoul, Marsanne, and Roussanne grapes, along with some significant rosés, make for versatile food-pairing options.

MEAT

The full-bodied, sometimes-rustic spice from Northern Rhône Syrah or the ripe, intensity of Châteauneuf-du-Pape handle heavier flavors of grilled steak, summer sausages, barbecue, and smoked game. A plate of charcuterie would be completely at home with a glass of Rhône Tavel rosé.

CHEESE

Tourmalet, from sheep's milk, spotlights Condrieu's affinity for remarkable cheese. Aged Gouda, Parmesan, and cheddar are solid picks for pairing with the bold flavors of Rhône reds. Dry and crisp Tavel rosé brings both fruit and lively acidity, perfect for a variety of cheeses.

POULTRY

Grenache's dark fruit character works well with roasted duck, turkey, or braised chicken. The full-bodied, bright fruit of Marsanne and Roussanne, found in Hermitage Blanc or Crozes-Hermitage Blanc, make perfect partners for chicken in cream sauce or topped with a ginger-citrus marinade.

DESSERT

When it comes to the Rhône's dessert wines, look for one of the region's Muscat-based sweet, fortified wines, the most famous of which is Beaumes-de-Venise. These wines complement the sweet spectrum, from citrus desserts to berry tarts.

SEAFOOD

Lobster, smoked fish, shrimp, scallops, and soft-shell crab all provide delicious pairing opportunities for the engaging aromatics and full-throttle, juicy stone fruit flavors of Condrieu. (The grape is Viognier.) Or try a heady white Northern Rhône blend for all things seafood.

PRODUCERS TO TRY

Chapoutier, Delas, Dubœuf, Grand Veneur, Guigal, Jaboulet, La Colliere, Perrin, Paul Autard

LOIRE VALLEY, FRANCE

The Region and the Wine

Stretching from the Atlantic coast to central France, the cool climate of the Loire Valley tracks the Loire River for more than 600 miles (965.5 kilometers). Red, white, rosé, sparkling, dry, off-dry, and sweet wines may be found in the close to 200,000 acres (80,937 hectares) of planted vineyards, but the Loire Valley's international reputation has been built on its stellar whites. Sauvignon Blanc, Chenin Blanc, and Muscadet grapes claim the spotlight, with the key regions of Sancerre and Pouilly-Fumé producing world-class wines made from 100 percent Sauvignon Blanc grapes.

■ Loire Valley region

The Loire Valley is the largest white wine–producing region in France, with the soil and microclimates well suited to growing remarkable grapes. From the medium-bodied, well-concentrated wines of Pouilly-Fumé and Sancerre to the dry but lighter-bodied styles of Muscadet, to the full spectrum of expression found in Vouvray, Loire Valley wines bring an elegance and character complexity that's second to none. Wines from Vouvray are made from 100 percent Chenin Blanc grapes and range from dry (sec), to off-dry (demi-sec), all the way to sweet dessert wines (*moelleux*). The sweet-styled whites are capable of aging for decades, thanks to the wine's innate acidity.

Selecting and Pairing

The extremely food-friendly nature of the Loire's diverse mix of wines and subsequent styles make them a pleasure to pair. Known for its collection of white wines showcasing pure fruit character and bracing acidity, the Loire also offers the fresh flavors of Chinon, a red wine made from Cabernet Franc.

MEAT

Red Chinon wine (made from Cabernet Franc in the town of Chinon) is the Loire's answer to many hearty meat dishes. From roasted lamb chops to baked ham and smoked game, the raspberry fruit flavors, medium body, and well-honed acidity of the region's best Chinon make for very versatile food pairing.

CHEESE

The tangy, pungent aromas and flavors of Sauvignon Blanc made from Pouilly-Fumé and Sancerre provide an incomparable match to the regional flare of the piquant flavors found in local goat cheese (*chèvre*). Likewise, the heavier, sweet-styled wines of Vouvray make a worthy match for various forms of blue cheese.

POULTRY

The heavier poultry picks of duck, pheasant, and dark turkey meat call for the flavor profile of a Loire Valley Chinon. However, the white meat chicken–inspired dishes are the perfect pairing for the pure fruit and acidity of Sancerre and Pouilly-Fumé's crisp, refreshing Sauvignon Blanc grapes.

DESSERT

Fresh fruit–themed desserts made with an assortment of berries or apple- and pear-inspired pastries are an ideal match for the honeyed nuances of Vouvray's sweeter side, labeled *moelleux* and *doux*.

SEAFOOD

Fresh seafood has been paired with the ultra-dry, mineral-driven notes of the Loire Valley's Muscadet wines for centuries. Shellfish is the perfect accompaniment to a well-chilled glass of Muscadet, as is the regional freshwater fish dish smothered in rich butter sauce, *beurre blanc*. For fans of smoked salmon, the clear-cut intensity of a Pouilly-Fumé pairs with the oily nature of the fish brilliantly.

❧ PRODUCERS TO TRY ❧

Champalou, Château de Sancerre, Château Moncontour, Clos de Nouys, Closel, Ladoucette, Pascal Jolivet, Sauvion

ALSACE, FRANCE

The Region and the Wine

Tucked in the northeast corner of France along the German border, Alsace is white wine country, with Riesling, Gewürztraminer, Pinot Blanc, and Pinot Gris accounting for close to 80 percent of the region's ultra-dry white wine. Happily, the regional wines are typically labeled by grape varietal as well as region—one of the few places in France that makes label layout easy on the New World consumer.

The 70-mile (112.5-kilometer) stretch of Alsace is barricaded by the Vosges Mountain range to the west and the Rhine River to the east, forming a buffered zone with dry, sunny days; cool nights; and optimal grape-growing conditions. Many of the grapes, producers, and even the fluted green bottle styles of Alsace speak of the overtly German influence—a result, no doubt, of the region changing hands between Germany and France many times over the last three centuries. However, whereas German wines tend to maintain a sweet aspect, with often lower alcohol levels, the Alsatians convert every last bit of sugar into alcohol, resulting in wines that are characteristically dry, boast higher levels of alcohol, and are fuller bodied.

Alsace region

Selecting and Pairing

Alsace is one of the world's foodie hotspots. From bakeries full of French pastries and assorted breads, to farm-fresh fruits and veggies, to sausage and sauerkraut, to pies stuffed with a medley of meat or escargot doused in butter, the wines of Alsace are well suited to a variety of foods, flavors, and cuisine.

MEAT

Alsatian fare often mirrors hearty German sausage and sauerkraut dishes. This is where the fuller-bodied, well-concentrated whites shine the brightest. The intense flavors of an Alsatian Riesling or the rich minerality of a Pinot Gris bring out the best in a variety of beef, game, pork, and lamb.

CHEESE

The fragrant, semisoft local Muenster cheese provides a delicious complement to the intense, spicy aromatic character and lychee-infused flavors of the regional Gewürztraminer. The pungent aromas of blue cheese also marry well with the fruit-forward frame of a Gewürztraminer.

POULTRY

All Alsatian whites are made for poultry. If it's foie gras, opt for the exotic aromatics of Gewürztraminer. In the case of chicken set to an Asian theme, reach for either Pinot Blanc or a Riesling. Chicken curry calls for the bright apple appeal of Pinot Gris.

DESSERT

A late harvest (labeled *Vendanges Tardives*) Pinot Gris, Gewürztraminer, or Riesling is sheer opulence when sipped as dessert itself or served as a dramatic partner in the presence of bread pudding, crème brûlée, or fresh fruit tarts.

SEAFOOD

Riesling is the wine of choice for seafood. From trout to salmon and sushi to shellfish, Riesling's fresh acidity and crisp, fruit-forward flavors make a classic pairing for the various weights, flavors, and textures found in raw and seasoned seafood.

PRODUCERS TO TRY

Hugel,
Lucien Albrecht (Wolfberger),
Marcel Deiss,
Trimbach, Zind-Humbrecht

THE REST OF FRANCE: LANGUEDOC-ROUSSILLON, PROVENCE

The Region and the Wine

The collective AOC wine regions of the Languedoc-Roussillon (often shortened to just Languedoc) and Provence wrap around the sunny, southern semicircle of the Mediterranean Sea. They share many of the same red grape varietals—mainly Mourvèdre, Grenache, Cinsault, Carignan, and Syrah—with some regional overlap in their whites—Chardonnay, Marsanne, Viognier, and Sauvignon Blanc.

With the great diversity of grapes grown; plenty of sunshine to fully ripen the fruit; and a reputation for producing easygoing, everyday wines at budget-friendly prices, the wines of Languedoc and Provence are enjoying their own international renaissance as savvy consumers search for food-friendly wines from the adventurous, lesser-known regions of France. The charming wines of Languedoc-Roussillon fall into two categories: Appellation-designated wines carrying the regional place names (Banyuls, Corbières, and Saint-Chinian, for example) and the Vins de Pays d'Oc, which can be loosely labeled according to grape type and sourced from all over the region.

■ Languedoc-Roussillon region
■ Provence region

Selecting and Pairing

Rustic, herbaceous, food-ready red wines constitute the majority of the wines from Languedoc, while Provence is known worldwide for bold, earthbound, yet fruit-filled rosés that provide a dynamic backdrop for the feisty, fresh flavors of Provencal cuisine.

MEAT

Braised beef, savory sausage stews, and rosemary-laden lamb shanks call for Languedoc's spicy, herbal tones found in the regional red blends, which are often based on the energetic grapes of Mourvèdre, Grenache, Cinsault, Carignan, and Syrah.

CHEESE

The ultra-sweet fortified flavor of Banyuls creates a consuming synergy with the stout, pungent flavors of Roquefort and Stilton blue cheese.

DESSERT

The famous sweet fortified red wines based on the Grenache grape from Banyuls offer a delectable dessert choice for dark chocolate or the creamy, coffee themes of tiramisu.

POULTRY

Regional rosé is a natural, versatile pairing for herbs de Provence–coated poultry or the tangy, Mediterranean medley of tomatoes, garlic, olives, and onion (dubbed *poulet Provencal*) that saturate braised or roasted chicken.

SEAFOOD

Provencal rosés are delicious with the seafood smorgasbord of bouillabaisse or the monkfish-inspired, garlic-heavy flavors of the local *bourride.* The smoked mussel dishes, *brasucade,* of Languedoc pair well with the spicy reds, and Languedoc's Picpoul is a top pick for shellfish.

PRODUCERS TO TRY

Chateâu d'Aussières, Chateâu d'Esclans, Domaine Clavel, Domaine Mirabeau, Domaine Tempier, Gérard Bertrand, Jean-Louis Denois, Les Close Perdus

WINES OF ITALY

With 20 distinguished wine-growing zones, estimates ranging from 800 to 2,000 grape varieties grown, and centuries of wine-making experience, Italy can be an exciting, albeit intimidating, wine venture for beginners and experts alike. By focusing in on some of Italy's most important wine-growing zones—Tuscany, Piedmont, and the Tre Venezie—you'll be ready to tackle a broad spectrum of food-friendly, regionally diverse Italian wines right from the start.

Historically, Italy has embraced its traditional wine-making roots and courted vines and subsequent wines that offer a particular affinity and pairing prowess with the fresh, seasonal flavors of the regional fare. Yet the last two decades have seen winemakers from all over the country take a dualistic approach to wine-making. Embracing tradition and relying on technology when relevant, vintners have cultivated a deep respect for the regional wines and experimented with different grapes, vinification techniques, and cutting-edge technology. It was this willingness to think, grow, and ferment outside of the DOC-induced box that led to the distinguished proprietary wines dubbed "Super Tuscans." From classic Chianti to the bold character of Barolo and the lighter styles of Pinot Grigio (same grape as Pinot Gris in France) and popular Prosecco, Italy's wines cover an extraordinary range of palate profiles, wine styles, and price points.

ITALIAN WINE CLASSIFICATIONS

Italian wines are most commonly named by region—typically a town, village, or district name such as Chianti or Barolo. They may also be named by grape variety, although this isn't nearly as common. Examples include Pinot Grigio, Sangiovese, Prosecco, and Nebbiolo. At times, labels highlight both the grape variety and the region, as is the case with Brunello di Montalcino. (Although this one gets tricky because the Brunello grape is a special strain of Sangiovese—not that a wine newbie would know that—and Montalcino is a prestigious hilltop village in Tuscany.) Finally, Italian wines can carry a creative proprietary name. The ones that often come to mind are pricey Super Tuscans with titles like Sassicaia or Ornellaia.

Italian wine laws establish a loose hierarchy for the country's wine quality levels. Originally intended to be a significant barometer of quality, these classifications influence the wine-making process in numerous capacities, yet they cannot be viewed as the ultimate measure of a wine. Delicious exceptions exist on both sides of the DOC designation. Some Italian table wines surprise with their ability to bring out the very best in a local dish, while other DOCG wines are completely lacking in presence and pairing ability. Only personal palate experience; adventurous food pairings; and attention to producers, regions, and wines that appeal to individual wine preferences can truly give context for quality in Italy's vast wine landscape.

Let's take a closer look at Italian wine classifications:

DOC (*Deonominazione di Origine Controllata*) label designations indicate that a wine has met specific regulations on what types of grapes may be grown and where, with detailed standards for vineyard yields; vinification processes that include minimum levels of alcohol; along with bottle and barrel aging requirements; and specific vineyard practices for trellising, pruning, and irrigation. Currently, Italy is home to more than 300 DOC wines.

DOCG (*Denominazione di Origine Garantita*) labeled wines must meet all the requirements for DOC classification but also carry an additional guarantee (*garantita*) that they adhere to even higher standards for aging and vineyard yields. It's worth noting that Piedmont and Tuscany carry a large portion of the DOCG wines—an indicator that although the DOCG labels are not a perfect system for pinpointing quality, they can certainly point consumers in the right direction.

IGT (*Indicazione Geografica Tipica*) loosely translates as "wines that are typical for their geographic region." Ironically, the IGT wine designation was the Italian government's response to the high-dollar, high-profile Super Tuscans being produced by gutsy winemakers completely outside DOC regulations. Technically, IGT wines are one step up from the Vino da Tavola (table wine) designation.

VdT (*Vino da Tavola*) wines can be made from any grape in any location within Italy. These everyday wines grace local tables as readily as a jug of water, if not more so. This group includes any wines that do not carry the DOC, DOCG, or IGT label designation.

RUFFINO

ORVIETO
DENOMINAZIONE DI ORIGINE
CONTROLLATA
CLASSICO
ABBOCCATO

Questo vino è prodotto con uve coltivate nei
vigneti intorno all'antica città di Orvieto,
terra ricca di tradizione vitivinicola.

VILLA DI
CAMPOBELLO
Dal 1860

CHIANTI
Denominazione di Origine Controllata e Garantita

ALC. 12% BY VOL   PRODUCT OF ITALY   RED WINE

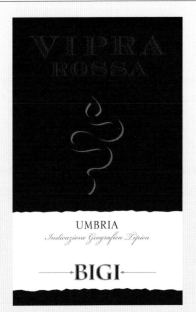

VIPRA
ROSSA

UMBRIA
Indicazione Geografica Tipica

BIGI

PIEDMONT, ITALY

The Region and the Wine

Rounding out the northwest corner of Italy's geographical boot, Piedmont is red wine and white truffle country. Literally meaning "foot of the mountain," Piedmont's rolling hillsides play cordial host to the vineyard landscape. The region's three dominating red wine grape varietals are Dolcetto, Barbera, and Nebbiolo, in order of ascending intensity. Nebbiolo is the building block for the legendary big, burly red wines of Barolo and Barbaresco (the names of the key towns that produce the renowned wines).

Piedmont region

The wines of Barolo and Barbaresco are powerful, highly concentrated, surprisingly complex reds capable of aging for decades, and they typically cost a pretty penny. When consumed young, they'll shock with harsh tannins and incredibly astringent qualities. However, when given some significant cellar time to tame the tannins and a decent decanting, these wild reds willingly untangle themselves into delightful layers of earth and smoke, cherry and chocolate, pumpkin spice and tobacco leaves, exuding power and purpose, folding the hearty flavors of truffles and cream sauce along with the meaty medleys of *bollito misto* (a traditional stew of boiled meat), braised game, and osso buco (veal shank) into rich, food-friendly companions.

Selecting and Pairing

Dolcetto and Barbera are everyday red wines perfect for pairing with risotto and mushroom sauce, but Barolo and Barbaresco are reserved for special occasions. The lighter frizzante, Moscato d'Asti, suits brunch dishes like berry-based recipes or fruit-filled crepes.

MEAT

The region's rustic Barbera reds are a solid pairing for everything from pizza to spaghetti and meatballs. The bold flavors, firm tannins, and stout structure of the big reds, Barolo and Barbaresco, are reserved for substantial cuts of meat that demand power and concentration.

CHEESE

The full flavors of Parmesan and Pecorino Romano bring out the best of both Barolo and Barbaresco. For a mild, semisoft cheese like fontina or the creamy notes of a young Gouda, the less-intense palate profile of Barbaresco brings out the ultimate elegance in both the cheese and the wine.

POULTRY

The lighter body and fruity nature of Dolcetto marry well with baked chicken, chicken cacciatore, and roasted game birds. Often packed in tomato sauce, garlic, and olive oil or pesto, Piedmontese poultry prefers the soft, subtle tannins and moderate acidity of Dolcetto reds.

DESSERT

Moscato d'Asti is Piedmont's perfect dessert partner. Low in alcohol, semisparkling, and off-dry in style, Moscato (the grape) d'Asti (from the town of Asti) elevates everything from meringues, to soufflés, to fresh berries topped with mascarpone cream, to the decadence of cheesecake.

SEAFOOD

Gavi is Piedmont's ultra-dry, highly acidic, citrusy wine from the town of the same name in southern Piedmont. Delicious with fried fish, calamari, sea bass, or the local cioppino or bouillabaisse stews, Gavi's crisp minerality cuts the fat and highlights the flavor of top seafood picks.

PRODUCERS TO TRY

Castello del Poggio, Ceretto, Gaja, Giacomo Conterno, Luciano Sandrone, Marcarini, Marchesi di Barolo, Mascarello, Poggio le Coste, Produttori del Barbaresco, Saracco, Vietti

TUSCANY, ITALY

The Region and the Wine

Tuscany is the heartbeat of Italy, centrally located and capturing scenic seaports, sun-drenched hillsides bursting with grapevines and olive groves, and dotted with medieval monuments surrounded by renowned Renaissance art. The red wine grape that has placed Tuscany on the world's wine map is Sangiovese, sold most often as Chianti.

Tuscany region

Chianti, the largest of Tuscany's wine regions, is composed of seven individual growing districts, all of which rely heavily on the Sangiovese grape and are regulated by DOCG wine laws. Wines labeled Chianti must contain at least 80 percent Sangiovese and may include a district name. Chianti is light to medium bodied, with rustic flavors, cherry fruit, and awesome food-friendly acidity, perfect for tomato sauce, olive oil, and pesto. The best districts are labeled under the designate Chianti Classico, typically a step up in price and quality from generic Chianti, using grapes from the hillsides of the better growing regions with a bit more body and spicy complexity. Chianti Riserva wines are also from the Chianti Classico region and have seen at least 2 years of oak aging, bringing smoky nuances with more concentration and intensity. *Super Tuscans* refers to Tuscany's pricey, bold reds produced outside the DOC-designated growing zones often using Sangiovese and the best of Bordeaux in the blend. Significant proportions of Cabernet Sauvignon, Cabernet Franc, or Merlot results in powerfully built red wines with solid structure, rich fruit, and well-managed acidity. Brunello (the grape) di Montalcino (from region of Montalcino) is an important DOCG in Tuscany. The Montalcino region is just south of Chianti and makes extremely full-bodied, age-worthy red wines from a specific strain of the Sangiovese grape called Brunello.

Selecting and Pairing

Italians enjoy a serious affection for both food and wine. Tuscany's wines are all food wines, with the lighter-bodied Chiantis pairing up well with lighter dishes and the fuller-bodied Chianti Classico labels working well with heavier fare.

MEAT

The fuller palate profiles, concentrated cherry flavors, and firmer tannins found in Chianti Classico, Riserva, Brunello di Montalcino, or Super Tuscans make a natural pairing for pastas with tomato and meat sauce (Tuscan Bolognese or lasagna), as well as the classic Tuscan white bean, kale, and sausage soup, or braised beef, various spicy sausages, smoked meats, and grilled game.

CHEESE

Italian mozzarella, Parmesan, provolone, and pecorino pair well with the region's most well-known wines. Chiantis—including many Super Tuscans—bring the right balance of acidity, fruit, and structure to the soft, creamy textures of mozzarella and can tone down the sharp, bitter edges of aged Parmesan.

POULTRY

Roasted chicken, quail, and turkey bring out the best of the black cherry and earthy nature often found in Chianti. Tuscany's Vernaccia di San Gimignano, the region's dry, aromatic white wine made from Vernaccia grapes, is a lively white wine choice for pairing with herb-seasoned poultry dishes.

DESSERT

Vin Santo is Tuscany's off-dry to sweet solution to the dessert wine dilemma. Delicious with desserts inspired by hazelnuts, pecans, almonds, and chocolate or berry-infused pastries, Vin Santo's sweeter side, with rich caramel flavors and silky textures, are also perfectly capable of standing alone as dessert itself.

SEAFOOD

The medium-bodied, slightly citrus character of Vernaccia di San Gimignano rises to the top of the pairing protocol for fried fish, lobster bisque, and shellfish.

PRODUCERS TO TRY

Antinori, Castello Banfi, Frescobaldi, Folonari, Fontodi, Luce della Vite, Ruffino

TRE VENEZIE, ITALY

The Region and the Wine

The large wine region of northeast Italy referred to collectively as *Tre Venezie* (the "three Venices") is comprised of several key wine zones: Veneto, Trentino-Alto Adige, and Friuli.

Veneto's wine region covers the fertile flatlands spiraling out from the seaside canals of Venice to the foothills of Verona, and the northwest border of Friuli and all of Trentino-Alto Adige are hemmed in by the Dolomites, a section of the eastern Italian Alps. Italy's lively whites hail from this northeastern pocket, especially from Alto Adige and Friuli, with Chardonnay, Pinot Grigio, Pinot Bianco, Müller-Thurgau, and Gewürztraminer taking the regional white grape spotlight. Reds range from the local favorites of Lagrein and Schiavo in Trentino and Alto Adige to the international varieties of Cabernet Sauvignon, Cabernet Franc, and Merlot.

■ Tre Venezie region

In Veneto, the dry white wine Soave and the popular sparkling wine Prosecco shine alongside three key reds made predominantly from the native Corvina grape. Bardolino is the Veneto's lightest-bodied red; followed by the medium-bodied, intense cherry aromatics of Valopolicella; and the dry, full-bodied, high-powered (and high-alcohol) Amarone reds.

The largely German-speaking northern region of Alto Adige, also called Südtirol or South Tyrol, has a heavy Austrian influence based on borders and history. Trentino leans heavily on Italian culture and is known for making lovely sparkling wines.

Selecting and Pairing

The Tre Venezie food scene is varied, with Austrian and Mediterranean influences impacting the cuisine. Various forms of cured ham, prosciutto in Friuli, and speck in Alto Adige are regular regional pairing partners.

MEAT

The local Lagrein, with its medium body, moderate acidity, and fresh berry aromatics, pairs well with the meat-themed flavors of the northeast. The Veneto's Valpolicella is a delight with sausage and polenta with meat sauces and handles the earthy flavors of meat and veggie dishes well. Pinot Grigio is a refreshing option with prosciutto and speck. For the rich, well-marbled cuts of meat, try Amarone.

CHEESE

A range of mild cheeses like mozzarella or Muenster to the sharper flavors of smoked cheeses demand Pinot Grigio. For easy pairing, pick a Lagrein or a Valpolicella Classico with full-favored, well-aged cheese choices. Or try a bold Amarone and aged Parmesan.

DESSERT

The sweet, concentrated flavor's of a late harvest Gewürztraminer from Alto Adige serves as an elegant pairing for the local apple-themed desserts.

POULTRY

Chicken, quail, and turkey are prime poultry picks for the medium-body and fresh cherry flavors of Valpolicella. The high acidity profiles of Soave and Pinot Grigio bring out the best in roasted chicken with pesto or other herb and garlic influences.

SEAFOOD

Lighter fish fare, sautéed shrimp, lobster, crab, clams, and calamari are well matched with the medium-bodied, dry styles of regional Pinot Grigio and Pinot Bianco. Tossing the seafood in lemon sauce just improves the marriage.

❧ PRODUCERS TO TRY ❧

In Vento: Anselmi, Allegrini, Bolla, Cantina del Castello, Cesari, Gini, Masi

In Trentino-Alto Adige and Friuli: Abbazia di Novacella, Alois Lageder, Ecco Domani, Ferrari, Folonari, Elena Walch, Il Poggione, J. Hofstätter, Jermann, Santa Margherita, Terlano, Tiefenbrunner, Tramin

ATLANTIC OCEAN

FRANCE

Bilbao ○

Barcelona ○

○ MADRID

PORTUGAL   SPAIN

Mediterranean
Sea

Wines of Spain

Spain is the third-largest wine-producing country in the world, after the heavy hitters of France and Italy, yet it boasts the most land dedicated exclusively to vineyards with close to 3 million acres (1,214,056 hectares) under vine. Grapes have been grown in Spain for thousands of years, with quality wines coming on strong in the last several decades. Spurred on by Spain's inclusion in the European Union, many Spanish wine regions have seen a focus on investment, technology, and irrigation techniques that have culminated in the growth and production of stellar grapes and spectacular wines. Spanish wines promise impressive value at virtually every price point, with built-in aging requirements and quality in high gear. These are wines to keep your eye on.

Spain has an interesting edge over many New World wines, with wine laws that set minimum aging standards prior to release. Even the entry-level Rioja Crianzas bring a significant age advantage over many of the bottle-and-release wines found in most New World growing regions.

Although Spain maintains the cultivation of more than 600 grape varieties, the most common can be boiled down into a dozen key players. Red grapes reign in the regions of Rioja, Ribera del Duero, and Priorat, with Tempranillo and Garnacha emerging as the leading varieties. White wines win the crown in the coastal climate of Rías Biaxas, with Albariño the most well known, and a healthy mix of both white and red grapes are grown in Penedès, the country's Cava capital. Spain's fortified crown jewel, Sherry, hails from the country's sun-kissed southern tip and is made predominately from the Palomino and Pedro Ximénez grapes. (More on Sherry later.)

Spain offers an amazing array of versatile, food-friendly wines and unique regional styles, from the lively sparkling Cavas to the rich, robust Rioja reds, to the crisp, refreshing flavors of Albariño, ensuring plenty is available to pair with local dishes and foodie trends.

Spanish Wine Classifications

Essentially Spain is home to four major wine classifications—DOC, DO, Vino de le Tierra, and Vino de Mesa—which bear a striking resemblance to French and Italian wine regulations. Specific regions, like Rioja and Ribera del Duero, break down their wine classifications even further based on the grape's quality and total aging time.

Here are the Spanish wine classifications:

DOC (*Denominaciones de Origen Calificada*) wines tend to be Spain's top bottles, with strict requirements for vineyard management, wine-making practices, and both barrel and bottle aging. Specific grapes grown in specific places with an emphasis on low vineyard yields are all dictated by the DOC regulations. So far, only the regions of Rioja and Priorat carry the exclusive rights to the DOC status.

DO (*Denominaciones de Origen*) wines are labeled to highlight the specific place name the wine comes from, with regulations dictating allowable grapes and critical geographical boundaries for that growing area.

VT or VdlT (*Vino de le Tierra*), meaning "wine of the country," are on the same playing field as the French "country wines," with grapes sourced at large from a singular region.

Vino de Mesa wines are the epitome of table wines, made from grapes grown from anywhere in Spain.

WINE STYLES

Oak is a big deal and big money in Spain. The more time a wine has spent in oak and the newer the oak, the pricier the wine. The highest-quality grapes generally earn more time in oak.

Rioja and Ribera del Duero take their oak seriously, resulting in a clever system for communicating the wine and oak relationship for the various styles and subsequent prices and quality levels of these regional wines:

Joven, meaning "young," indicates the wine has either spent no time in oak or just a smidge and should be consumed sooner rather than later. Fruit is the primary focus here.

Crianza wines have met minimum oak requirements and tend toward a light to medium body with fresh fruit flavors, some spicy oak-derived character, and plenty of points for value pricing.

Reserva wines must be aged 3 years in oak for red and 2 years for whites (a mix of barrel and bottle aging) and bring more body, complexity, and intensity thanks to the extended oak influence.

Gran Reserva wines are only made in the best vintages and have seen significant aging prior to release: 5 or more years for reds (with a 2-year minimum in oak barrels) and 4 years for whites (with a 6-month minimum in oak). Expect layers of intriguing aromas and supple tannins in the reds as well as greater overall concentration.

Spanish Wine Label Tips

Designate	What It Means
Tinto	Red wine; labels read *Vino Tinto* for "red wine"
Blanco	White wine; labels read *Vino Blanco* for "white wine"
Rosado	Rosé wine
Pago	Vineyard; labels might read *Vino de Pago,* indicating a single vineyard wine
Cosecha	Vintage or harvest
Bodega	Generic term meaning "winery"

Rioja, Spain

The Region and the Wine

Rioja wraps itself along the Ebro River in north central Spain, with the coastal Cantabrian Mountains providing a protective barrier against the Atlantic. Again, place is the theme in Spain, not grapes, and *Rioja* is common vernacular for discussing the distinctive regional reds. Rioja is divided into three subregions: the cool-climate districts of Rioja Alta and Rioja Alavesa and the balmy Rioja Baja, all of which take their own twist on the Tempranillo grape. Rioja Alta is perched at a decent altitude, making wines with solid acidity, impressive color extraction, and moderate alcohol levels. These tend to be the better wines from Rioja.

Rioja is best known for its world-class red wines made predominantly from the spicy, ripe cherry flavors of Tempranillo and the supporting character of Garnacha. Historically, Rioja's red wines have earned a reputation for being spicy and supple with layers of oak-tamed fruit. Today, many bodegas are experimenting with a fresh, fruitier style of red wine with less oak influence. Rioja's white wines play a small regional role and are based on the Viura grape, often blended with Malvasia or Garnacha Blanca.

■ Rioja region

PRODUCERS TO TRY

Baron de Ley, Montecillo, Muga, Campo Viejo, El Coto, Finca Allende, La Rioja Alta, Marqués de Cáceres, Marqués de Murrieta, Museum

RIBERA DEL DUERO, SPAIN

The Region and the Wine

Dotted with impressive medieval castles, Ribera del Duero hugs both sides of the Duero River and is located less than 100 miles (161 kilometers) north of Madrid. Covering hillsides and high plateaus, Ribera del Duero's vineyards are exposed to hot, arid summers with blazing daytime heat contrasted by cool evenings. This temperature swing optimizes the grapes' acidity, aromas, and overall color components.

Red wines dictate in Ribera del Duero, and Tempranillo prevails as the primary regional red wine grape; however, it often goes by the hometown alias of *Tinto Fino*. This dry, rugged region is known for taking Tempranillo to remarkable levels of quality and concentration. Full-bodied wines of exceptional character, with the dark fruited flavors of plum, blackberry, and black currant wrapped in super-supple tannins, have received international attention in recent years. Other common regional red wine varieties include Garnacha (especially popular in local rosados) and Cabernet Sauvignon. Thick, gnarled 50- to 80-year-old vines protruding from the dusty hillsides of the Ribera del Duero are not uncommon. In fact, many of these old vines are the backbone of the area's robust red wines.

■ Ribera del Duero region

PRODUCERS TO TRY

Abadia Retuerta, Bodegas Alejandro Fernandez, Bodegas Reyes, Bodegas Vega Sicilia, Condado de Haza, La Pesquera, Viña Sastre

PRIORATO, SPAIN

The Region and the Wine

Two hours south of Barcelona, the rugged wine-growing region of Priorato produces high-octane reds. The old, gnarled vines of Garnacha and Cariñena (Grenache and Carignan in France) overshadow the steep hillsides and terraced vineyards of the harsh, infertile landscape. Recognized as one of two high-caliber DOC wine regions in Spain, Priorato produces powerful red wines that maintain incredible intensity, massive tannins, and high alcohol levels, all from the very low yields of very old vines.

Residing in Spain's Catalonia region, locals here refer to their wine region as *Priorat* while the rest of the country calls it *Priorato*—expect to see *Priorat* on labels. Given the harsh growing conditions, sizzling summers, and bitter winters, the largely hand-harvested, steep-sloped vineyards must struggle against a difficult climate and inhospitable, slate-filled soils. It's no surprise vineyard yields are quite low, and the fruit arrives deeply concentrated. At times you may find the native Cariñena and Garnacha grapes blended with Cabernet Sauvignon for added structure and longevity in addition to small percentages of Syrah and Merlot in the mix—although again, labels will indicate Priorat, not the grapes. French oak is a popular choice for aging, while many other regions still rely heavily on American oak. Wines from Priorato tend to be pricey, in part because the region's production is relatively low, yet demand stays steady as the region's reputation continues to shine. Look for Priorato's high-powered, full-bodied red wines with plenty of inky, dark colors, concentrated fruit, and complex character.

Prioriato region

PRODUCERS TO TRY

Alvaro Palacios, Cellar Pasanau, Clos Erasmus, Clos Les Fites, Clos Martinet, Clos Mogador, Finca Dofi, Scala Dei

PENEDÈS, SPAIN

The Region and the Wine

Penedès is Cava country. Cava is the region's budget-friendly sparkling wine, made from several local grapes: Macabeo, Parellada, Xarel-lo, and more recently Chardonnay. Charming bubblies are Penedès' top ambassadors, but still wines are widely produced. Most are whites, but plenty of ambitious, fruit-forward reds are making their way to international markets. Located outside of Barcelona, Penedès borders the Mediterranean coastline and stretches through hilly terrain up toward the Catalan coastal mountain range. This mix of topography and soils makes for diverse grape-growing conditions teeming with countless microclimates.

Abundant sun, warm temperatures, and mild Mediterranean winters give Penedès plenty of vineyard support. The famous Spanish sparklers are created in the traditional Champagne method, performing the second fermentation, which creates the bubbles, in the bottle. Cava can range from bone dry (brut) to very sweet (dulce), although the brut tends to be the most popular. Regionally inspired tapas, fried foods (especially fish), and the famous Spanish Serrano ham all partner quite well with the engaging aromatics, citrus-infused flavors, and lively acidity of Cava.

■ Penedès region

PRODUCERS TO TRY

Castillo Perelada,
Codorníu, Freixenet,
Jaume Serra Cristalino,
Jean Leon, Segura Viudas,
Torres

Rías Baixas, Spain

The Region and the Wine

Tucked up in the northwest corner of the country along the Atlantic coast and sitting just north of Portugal, Rías Baixas is Spain's celebrated white wine country. Here, the Albariño grape produces distinct white wines full of fragrant floral aromas and pure fruit flavors. The region's seafood scene creates the perfect pairing partnership for Albariño's generous acidity and crisp, clean flavors. Rías Baixas wines are bone dry with a pale golden color, are highly aromatic, and embrace the fruit flavor range of green apple, pear, citrus, melon, peach, and apricot with undertones of warm spice and minerality.

The cool, maritime climate of the Rías Biaxas is at the mercy of the Atlantic Ocean, with wet and rainy conditions quite common. The climate influences the grapes in significant ways, ensuring plenty of bracing, food-friendly acidity. Virtually all the wines from Rías Baixas are designated as DO wines, and the Albariño grape name graces the regional labels, often a helpful switch from the place name labels that rule the rest of Spain. Recent decades have seen updates in vineyard practices and equipment modernization. The technology of temperature-controlled stainless-steel fermentation tanks has led this modernization charge, allowing for the full, pure fruit expression of Albariño to take shape. Affordable Albariño is intended to be consumed well chilled in its youth while the fruit is fresh and energetic.

■ Rías Baixas region

❧ Producers to Try ❧

Bodegas Martin Códax (and Burgáns), Bodegas Morgadio, Pazos de Lusco, Pazo de Señoráns, Vionta

Selecting and Pairing

For overall meat pairing flexibility, the first stop is Rioja. The various styles of Rioja's reds provide ambitious pairing complements for an impressive assortment of meat-based fare. Likewise, Rioja's whites offer plenty of versatility when partnered with poultry, paella, and more.

MEAT

From roasted lamb and pork dishes to beef stews with a mix of peppers, onions, garlic, and tomatoes, to beef-themed tapas or Spain's Serrano or *jamón ibérico* (ham), Rioja's reds make pleasing combinations. Lamb chops continue to be the regional choice for pairing with the full, earthy flavors of Ribera del Duero's reds.

CHEESE

The bright acidity of Spain's sparkling wine, Cava, and the dry to off-dry Sherries marry well with a slice of young, semisoft local Manchego. For reds, nothing compares to Rioja. Consider a well-aged Manchego with the structure and innate complexity of an oak-aged Rioja Reserva. For a sharp blue cheese, opt for a full-bodied Garnacha-based wine from Priorato.

POULTRY

The lighter, fruitier Crianza styles of Rioja's reds provide delicate pairings for poultry. Rioja's oak-driven reds handle the heavier flavors of game birds like pheasant or wild turkey exceptionally well. Spain's bubbly Cava is the perfect partner for fried or sautéed chicken recipes.

DESSERT

Spain's decadent desserts call for the rich, syrupy textures of a cream sherry or Pedro Ximénez (PX). Local favorites like flan or Catalan cream marry well with the rich palate profiles of Pedro Ximénez Sherry. Chocolatey desserts and pecan pie are also a delight with PX, while cheesecake is a heavenly match for the silky textures and similar flavors of a cream sherry.

SEAFOOD

Spain's best option for pairing with all seafood is Rías Baixas' favorite white, Albariño. The higher levels of acidity; crisp clean citrus, pear, and peach fruit flavors; and underlying minerality are outstanding alongside the delicate flavors of the local scallops and oysters; the oily nature of fresh fish or grilled octopus; or spicy clams, lobster with lemon butter, Cajun cuisine, and zesty ceviche.

Braga

Porto

PORTUGAL

SPAIN

*ATLANTIC
OCEAN*

LISBON

WINES OF PORTUGAL

Portugal hosts more than 250 different grape varieties, including familiar international grapes, packed into more than 20 wine-growing regions, with the top regions carrying the coveted DOC status. Holding the title for being the European Union's fifth-largest wine-producing country, Portugal weighs in with close to 70 percent of the country's wine falling into the distinctly red wine category. Cork is also big business in Portugal. The nation's cork forests, largely located in Alentejo, produce close to 50 percent of the world's raw cork material.

Portugal is best known for Port, the sweet fortified dessert wine, but the last decade has seen a significant focus on the region's up-and-coming red wine production. We'll save Port for later in the book when we look at fortified wines in more depth, but for now, it's interesting to note that many of the same native grape types traditionally used for making Port have been reassigned to influential roles in the country's red wine production. That's not to say regional white wines don't play a compelling part in Portugal's overall wine arena. They do. It's just that Portuguese reds have seen a surge in quality, consistency, and intensity, which is catching worldwide attention.

Wine laws in Portugal are similar to the rest of the European Union, with designations for Vinho de Mesa (table wine), Vinho Regional (VR; regional wine) or Protected Geographical Indication (IGP), and DOC wines (bottling strictly regulated wines from a specific region or place).

Wine labels in Portugal may come with a back-label seal of authenticity, indicating which region in Portugal the wine is from. Front label lingo to know: *quinta* refers to a wine estate, much like *chateau* in France or *bodegas* in Spain. *Tinto* means "tinted," or a wine with a colored tint, a red wine. Lastly, *vinho* is Portuguese for "wine."

Portuguese wines represent some of the world's best wine bargains, with terrific value whites and well-priced, ageable red blends.

PORTUGAL'S DOURO, DÃO, AND ALENTEJO REGIONS

The Region and the Wine

The Douro and the Dão, both in northern Portugal, are leading red wine–producing regions. The Douro is a warm, rugged river valley that follows the Douro River from Spain to the city of Porto, where it empties into the Atlantic. The dominant soil in the Douro is called *schist* and has a flinty character that comes from shale. Not necessarily "fertile" ground, schist forces the vines to struggle in their search for water and nutrients, creating wines that often exude deeper concentration and overall intensity.

Portugal's Dão region, again named for the main river, the Rio Dão, enjoys protection from significant heat and cooler evening temperatures, thanks to the Serra da Estrela mountain range.

- ■ Douro region
- ■ Dão region
- ■ Alentejo region

The large, arid region of Alentejo, located in Portugal's southern sphere and sharing a border with Spain and the Atlantic, has also contributed significantly to the country's red wine portfolio. The majority of the quality white wines come from the Minho district in the northwest, just across the border from Spain's white wine capital, Rías Baixas. Alvarinho (known and grown as Albariño in Rías Baixas) constitutes the stellar white wine grape used in crafting the leading Vinho Verde wines.

The major red wine grapes to be familiar with include Touriga Nacional and Aragonez, although many of Portugal's red wines are made with a blend of native grapes, often in a "field blend" or intentionally mixed in-house. The aromatic, tannic, and black-fruit character of Touriga Nacional splits its influence between Port production and many of Portugal's best red wines in the northern Douro region. Aragonez, same grape as Tempranillo in Spain, is frequently found in the robust red wines from Alentejo. The Minho region's key white wine, Vinho Verde, showcases fresh, white fruit character wrapped around a tight core of vibrant acidity. *Vinho Verde,* meaning "green wine," refers to the wine's youth, when it's best consumed. *Vinho Verde* most often refers to the young, lower-alcohol, sometimes-spritzy, fairly acidic white wine made predominately from the Alvariñho grape; however, it can also refer to young red wines.

Selecting and Pairing

Portugal cultivates hundreds of grape varieties, but that's not necessarily a priority when it comes to labeling laws. Place, not grape, graces the label. Blending is big business in Portuguese wines. Whether red, white, or fortified, the marriage of many grapes tends to find favor over bottling and showcasing a single grape. These wine blends often allow a wine to shine by amplifying a grape's good qualities and minimizing the rougher character components via the synergy of the blend. From the crisp, fresh, fruity notes of Vinho Verde whites to the rich, almost velvety character of the Douro's top reds, Portugal produces a wide range of palate appeal for pairing a broad spectrum of both local and international cuisine.

MEAT

The Douro reds, often dominated by the well-maintained tannin and dark fruit character of Touriga Nacional, create a dynamic food partnership with a range of red meat. From braised beef and wild game, to hamburgers, brats, meatloaf, and roasted lamb or pork roast, the sometimes-spicy, medium-bodied profile of both the Douro and Dão reds makes a convincing match with meat.

POULTRY

A variety of roasted duck and chicken recipes find a tasty companion with the lively character and fruit-centered aspects of a white wine blend from a local Alentejo. Or pick the fresh, vibrant profile of a recently released Vinho Verde to complement the savory flavors of a chicken salad topped with garlic and lemon dressing.

SEAFOOD

Just as Spain's sassy, white wine Albariño is made for seafood, so is Portugal's Alvarinho. Super fresh, often fruit-driven with heavy citrus and pear influences and an abundance of mouthwatering acidity, Alvarinho is the answer to the region's cod-inspired dishes (think *bacalhau*), fried calamari, garlic shrimp, and crab salad or seafood-themed stews.

CHEESE

The heavier red wines of the Douro and Dão are perfect for pairing with the mature flavors of aged cheddar or Gouda. The softer palate profiles of the red wines from the Alentejo region are delicious paired with raw goat's cheese, Manchego, or slightly smoked Gouda. However, the classic pairing of Port with the salty, pungent flavors of a full-flavored blue cheese (think Stilton or Roquefort) offers a remarkable wine and cheese alliance.

DESSERT

Port is perhaps the most-recognized dessert wine in Portugal and around the world. The sweet flavors, smooth textures, and nutty nuances of Port make it a dessert option in itself. However, when paired with dark chocolate or merged with pecans, walnuts, and almond dessert inspirations, Port creates an enduring palate impression. Fruit pies and pumpkin pies also play exceptionally well with a rich, full-bodied Port.

PRODUCERS TO TRY

Aveleda, Broadbent, Croft, Churchill's, Dow's, Esporão, Ferreira, Graham's, Luís Pato, Murças, Pinto, Quetzal, Roriz, Taylor Fladgate, Vallado, Warre's

WINES OF GERMANY

Known worldwide for setting the gold standard for the ultra-versatile Riesling grape, Germany is Europe's fourth-largest wine-producing country. The latest data indicates that 60 percent of Germany's wines are white and 40 percent are red. Some estimates note more than 100 different grape varieties are grown throughout German vineyards. However, in the country's export market, we see an overwhelming amount of white wine, with the fresh, food-friendly Riesling representing the top German export. Secondary white wine grapes grown in Germany include Müeller-Thurgau, Silvaner, Pinot Grigio, and Pinot Blanc. The chilly climate makes it more difficult to grow red wine grapes that require significant sun to fully ripen, but Germany has had success with Pinot Noir (known locally as Spätburgunder) and Dornfelder, although they're rarely seen outside the country.

Germany is home to 13 designated wine-making regions with 4 of them being of prime importance (and the names you'll see most often on German wine labels). The top regions are named after the major rivers that run through them, with the Mosel River (and region) being the most well known. The Rhine River (Rhein in German) winds through three other significant growing regions: Rheinhessen, Rheingau, and the Pfalz (which used to be called Rheinpfalz in keeping with the river-naming theme).

Germany's top regions all have cool growing climates, influential river topography, and steep hillside slopes covered in vines, but the unique variations in wines from vineyard to vineyard is astounding.

German Wine Classifications

Germany's wine classification system can be tricky to navigate, with quality levels determined by the ripeness of the grapes at harvest, not on vineyard locations. The more sun the vineyards catch, the riper the grapes will become; the riper the grapes, the more sugar they contain; the higher the sugar levels, the more alcohol that can be converted during fermentation. So a grape that reaches full maturity tends to make better wine. This is precisely why Germany has put a high priority on how ripe the grapes are at harvest time, especially considering the rather cool northerly climate. Capturing as much sun as possible prior to harvest is essential to ripening the region's grapes and making great wine. The dessert wines require a considerable amount of sun and warmer growing conditions; therefore, they're the only options in the best vintages and can claim quite a price when available.

Essentially, Germany has two distinct categories for quality. On the low end is *Tafelwein* (literally "table wine"), while on the other end of the spectrum is *Qualitätswein* (literally "quality wine"). Qualitätswein is then broken out into two additional tiers:

Qualitätswein bestimmter Anbaugebiete (QbA) is a wine produced in one of Germany's designated growing regions from grapes that were on the lower end of the ripeness scale at harvest time. These wines may undergo chaptalization (the addition of sugar during fermentation to increase the wine's final alcohol content) because the grapes were likely picked before they reached optimum maturity and sufficient levels of sugar.

Prädikatswein loosely translates as "quality wine of distinction." These wines are the top tier of the quality category for German wine, and the grapes must be ripe enough to forgo the fermentation process. The Prädikatswein category is further broken out into very specific definitions for ripeness levels (also indicative of quality and price), which appear on labels as Kabinett, Spätlese, Auslese, Beerenauslese, Trockenbeerenauslese, and Eiswein.

Here are the Prädikatswein wine categories listed in ascending order of the grape's ripeness at the time of harvest:

Kabinett These grapes are picked at relatively normal harvest times, have lighter bodies and lower alcohol levels, and are often made in an off-dry style. The wines are typically quite versatile, able to pair with a variety of foods, and carry reasonable prices.

Spätlese The grapes are picked later than the Kabinett grapes (*spät* means "late") and tend to enjoy a bit more body, concentration, and alcohol. They can be made in sweet or dry styles and are food-friendly wines with great acidity and fair pricing.

Auslese These grapes are picked cluster by cluster, and only the grapes exhibiting exceptional ripeness are selected for harvest. These are not annual occurrences but are reserved for the warmer vintages, when sufficient sun has graced the grapes with greater opportunity for maturity. The wines are full bodied and can run from dry, to off-dry, to sweet.

Beerenauslese (BA)   As we trend up the ripeness scale, Beerenauslese stands out because instead of harvesting the grapes by individual clusters, the late-harvest grapes are hand-picked by *the single grape*. Only the grapes that have been significantly affected by botrytis (also called noble rot) qualify for picking because of the intense concentration of sugar that results from the relationship between the friendly fungus and the grape. These wines are incredibly rich, dessert wines that are priced fairly high.

Trockenbeerenauslese (TBA)   *Trocken* means "dry" and refers to the individual, raisinlike grapes picked one at a time for the priciest dessert wines from Germany. Ultra-rich decadent wines with incredible intensity and honeyed flavors that often show lower levels of alcohol, TBAs are among the world's most extravagant dessert wines.

Eiswein   Meaning "ice wine," these wines are made from well-ripened grapes that have naturally frozen on the vine. The grapes are harvested and pressed while frozen, which allows the grape juice to easily separate from the frozen water drops. The result is a highly concentrated elixir with astounding acidity balanced by intense sweetness. These are understandably expensive wines, with so little juice to work with.

More recently, Germany has come up with another layer of quality in an attempt to communicate more clearly with consumers. The quality designation is **Verband Deutscher Qualitäts- und Prädikatsweingüter (VDP),** and it's seen on bottle capsules and labels as *VDP* with the logo of an eagle clutching a cluster of grapes. Currently less than 200 German wine estates carry the prestigious VDP designation.

GERMAN WINE LABELS

Germany's wine labels have a reputation for being complicated, with long words, ornate illustrations, and impressive designations, yet recent vintages have seen a slew of sleek labels offering only key information modeled after many New World labeling schemes.

The traditional labels can be quite eye-catching, with the estate's heritage, family crests, and brawny castles in the background, but once the place name data like vineyard, town, region, and quality levels take their place, the label can get overwhelming. So it's been with open arms that many of the clean, crisp, just-the-facts German labels have been welcomed onto international wine shelves.

German wines labels, whether old style or new, communicate several key pieces of information: the producer, the region and or vineyard, the vintage year, often the grape varietal, and the ripeness classification and quality level. Recently, Rieslings released in or imported to the United States have seen an increase in back bottle labels with a helpful Riesling Taste Profile scale, produced by the International Riesling Foundation, that shows a spectrum ranging from dry to sweet. Producers who choose to use this scale indicate where their wine falls on the sweet spectrum, helping take some of the guesswork out of the wine's style for eager consumers.

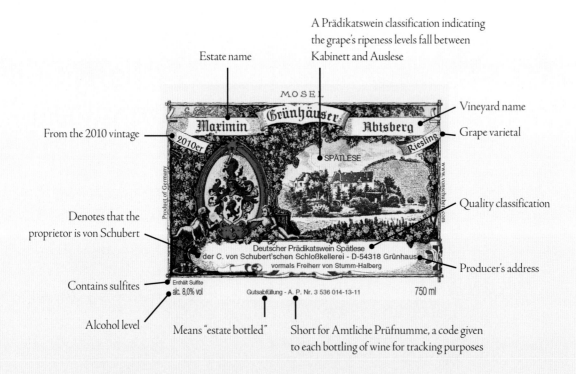

A Prädikatswein classification indicating the grape's ripeness levels fall between Kabinett and Auslese

Estate name

Vineyard name

Grape varietal

From the 2010 vintage

Denotes that the proprietor is von Schubert

Quality classification

Producer's address

Contains sulfites

Alcohol level

Means "estate bottled"

Short for Amtliche Prüfnumme, a code given to each bottling of wine for tracking purposes

German Wine Label Tips

Label Clues	What It Means
Trocken	Translates as "dry."
Halbtrocken	Indicates a wine is "half-dry" or off-dry.
Feinherb	A bottle term indicating off-dry.
Weingut	A wine estate that grows its own fruit.
Weinkellerei	A wine estate that sources its fruit from other growers.
Varietal	If a label names a grape type front and center, you know it contains at least 85 percent of that grape variety.
Ripeness levels	The grape's ripeness level (Kabinett, Spätlese, Auslese, Beerenauslese, Trockenbeerenauslese, and Eiswein) also appears on the label.
Big words or vintage years with random *-er* endings	Many German wine labels add an *-er* ending on vintage years and village names. This means "from" or "belongs to." For example, *2010er* is from the vintage year 2010, and *Wehlener* means the wine comes from the Mosel village of Wehlen, or as often seen on the label, *Wehlener Sonnenuhr*, meaning the grapes come from the vineyard of Sonnenuhr belonging to the village of Wehlen. This is similar to how Americans call someone from the south a Southerner or a person from a New England state a New Englander.

MOSEL, GERMANY

The Region and the Wine

The Mosel is Germany's most famous wine region, with some of the world's steepest vineyards clinging to the hillsides that trace the Mosel River. Mosel maintains an energetic reputation for producing world-class Rieslings that are are among the most delicate in style, with lighter bodies and an abundance of mouthwatering acidity and typically lower alcohol levels. Mosel Rieslings are generally bottled in tall, green, fluted bottles. (Neighboring Rheingau wines wear the traditional brown fluted bottles.)

Mosel's chilly climate takes plenty of credit for the dynamic balance of sugar and acidity in the region's best wines. Mosel Rieslings lean toward the fresh fruit flavor of apple and pear with occasional citrus notes. A distinct but delicious mineral-driven component prevails in many Mosel Rieslings, as the slate-filled soil leaves its distinguishing mark. The Mosel River Valley is teeming with countless high-quality producers that ship their extraordinarily food-friendly wines around the world.

Mosel region

GERMANY'S RHINE RIVER REGIONS

GERMANY

Rhine
○ Bonn
Mosel
Frankfurt am Main ○
Rhine
FRANCE
Stuttgart ○

- The Rheingau region
- The Rheinhessen region
- The Pfalz region

The Region and the Wine

Germany's Rhine (Rhein in German) River is the country's longest river and plays host to three of Germany's most significant wine regions: the Rheingau, Rheinhessen, and Pfalz. All three regions cultivate a variety of grapes, but Riesling still reigns throughout. Each wine-growing district along the Rhine puts its own spin on the Riesling grape, producing compelling variations based on sun exposure, hillside aspect, soil components, and individual vintner preferences.

The Rheingau constitutes one of the smaller growing regions and is marked by a series of sunny, south-facing slopes that maximize the grape's sun exposure and speed the ripening process. Riesling and Spätburgunder (a.k.a. Pinot Noir) are two of the top grape varieties. Rheingau Rieslings tend to be dry (*trocken*) and typically have a fuller body than their Mosel cousins, showing expressive apricot and peach fruit along with rich, round palate profiles. These drier Rieslings are balanced out by the region's exceptional sweet, late harvest dessert wines.

The Rheinhessen is Germany's largest wine-growing region and plays host to a variety of grapes with Riesling still prominent. Sylvaner also does exceptionally well here. With such a diverse growing zone, Rheinhessen wines run the gamut from high-quality, single-estate wines to the more generic Tafelwein status. This is when label knowledge and producer appeal comes in handy.

The Pfalz is a relatively warm region with a vibrant mix of soils and terroir. Many grapes are grown here (both red and white) and yield fuller bodies with less acidity and more finesse than many of their cooler-climate counterparts. The red Spätburgunder grape thrives here, thanks to the somewhat higher temperatures and increased access to sunlight.

Selecting and Pairing

Riesling rules when it comes to food-pairing versatility, with an innate ability to tone down the heat in spicy dishes with its semisweet to sweet palate profile. In Germany, many pair the wine with a variety of meat dishes, disregarding the rigid "white wines with white meat" idea. Rieslings work well with tough-to-pair foods like exotic salads marked with vinaigrette dressings or one of wine's worst enemies—asparagus (which is complemented by a dry or off-dry version). Riesling is a restaurant go-to wine when everyone at the table is ordering in different directions. It can handle starters, soups, salads, and a variety of veggies and then move right on to poultry and pork, and the drier styles with a bit more body can stretch to steak and even wild game. Spätburgunder and Dornfelder can hold their own when red wine options are desired.

MEAT

Just like Pinot Noir handles the full spectrum of red meats from steak, to pork, to lamb, to wild game, *Spätburgunder* (same grape, different name) pairs well with all these and also works with mushroom-based sauces—think stroganoff. Dornfelder is a very light-bodied red that runs a similar palate profile to Beaujolais' Gamay grape and pairs nicely with burgers and fries, roast beef or salami sandwiches, and grilled sausages. The light- to medium-bodied Riesling *Spätlese* is often the first choice for various pork cuts, and the sweeter styles provide a terrific pairing with some roasted meats.

POULTRY

Off-dry Rieslings partner well with nearly all poultry, but the hints of sweet bring out the best in roasted chicken, turkey, goose, and duck. The rich textures and flavors of foie gras are an amazing complement when paired with the similar textures and weight of Germany's sweet Beerenauslese or the intensity of an Eiswein. Spicy Asian themes like Kung Pao, Szechuan, or an Indian chicken curry are perfect partners for the slightly sweet profile of an off-dry Riesling *Spätlese.*

SEAFOOD

Freshwater fish from both the Mosel and Rhine provide plenty of Riesling pairing opportunities, and depending on the preparation of the fish, the off-dry (*halbtrocken*) style of a Riesling *Spätlese* complements a simple pan-seared, sautéed, herb-smothered, or smoked fish perfectly. Clams, mussels, oysters, and scallops are equal pairing partners for both dry and off-dry Rieslings.

CHEESE

Asiago, Brie, and fresh goat's cheese pair well with the dry and off-dry styles of Riesling Kabinett and Spätlese. For the bolder flavors of Stilton, opt for the rich dessert wines in the Beerenauslese or Trockenbeerenauslese categories.

DESSERT

Germany produces top-notch desserts and the wines to match. Consider fresh fruit desserts with apples, apricots, or peaches alongside the ultra-rich Beerenauslese or Trockenbeerenauslese. Custards, puddings, dessert crêpes, and caramel-inspired treats turn the dessert wine pairings into extraordinary partners.

PRODUCERS TO TRY

Dönnhoff, Dr. Loosen, Egon Müller, Fritz Haag, JJ Prüm, Josef Leitz, Maximin Grünhaus, Markus Huber, Markus Molitor, S.A. Prüm, Selbach-Oster, Schloss Johannisberg, Schloss Vollrads, Strub, Villa Wolf, Weil

WINES OF AUSTRIA

Austria has four key wine-growing regions: Lower Austria, Burgenland, Vienna, and Styria. The two major regions to be familiar with, however, are Lower Austria, which is actually in the northeastern corner along the fertile Danube River Valley, and Burgenland, which hugs the country's eastern border. Lower Austria is known for producing bright, full-bodied, bone-dry white wines, and the warmer growing conditions of Burgenland bring red wines with ripe, red fruit flavors and peppery aromas, along with ultra-rich dessert wines, to the table.

The last two decades have seen a significant leap in the quality of Austria's leading wines, thanks in part to a new generation of winemakers bringing innovation and modernization to the forefront of Austria's wine culture. Austria boasts 35 regulated grape varieties with more than 75 percent of them falling into the white wine category.

Austria's wines are often discussed in relation to neighboring German wines, with similar label lingo, usually with the same willowy wine bottle shapes, and grape varieties front and center on the labels. Austrian wines trend toward higher alcohol levels and more body than their German counterparts, partially due to the often warmer regional growing temperatures.

Delicious dessert exceptions withstanding, most Austrian table wines are made in a remarkably dry style.

AUSTRIAN WINE CLASSIFICATIONS

Austria shares many quality classification similarities with Germany's complex system based on a grape's level of ripeness and maintains some of the world's highest standards for quality classification. Sugar levels, alcohol levels, grape varieties, vineyard yields, fermentation processes, and more are all laid out by strict Austrian wine law. These standards also undergo rigid qualitative taste tests by a regulatory board to ensure accuracy and quality in addition to quantitative chemical analysis for every wine beyond the Tafelwein category.

Labels often wear grape names in addition to place names, and the bottle must contain a minimum of 85 percent of the featured grape. Same with the vintage year—at least 85 percent of the juice in the bottle must have come from the vintage year that adorns the label, leaving a little room for back-blending vintages on occasion.

Let's take a closer look at Austrian wine classifications:

Prädikatswein represents the top tier of quality for Austrian wines (and can be found on the label). These wines must adhere to a stringent set of rules in order to gain the prestigious classification. The wines undergo significant testing to guarantee the wine has, in fact, met all classification criteria. The Spätlese, Auslese, Beerenauslese, Eiswein, Schilfwein, Ausbruch, and Trockenbeerenauslese subcategories of wine (in loose order of relative sweetness from least sweet to most sweet) all fall under the Prädikatswein umbrella and are based on the grape's ripeness levels. The Trockenbeerenauslese wines, made from dried, raisinlike grapes, represent the most expensive dessert wines from Austria and boast incredible balance; intensity; and rich, honey character.

Qualitätswein means "quality wine," and represents the middle level—a step up in quality from Tafelwein, but not at the Prädikatswein tier. Qualitätswein must be produced in a designated wine region from one of the allowable grape varieties. This is an everyday wine with reasonable pricing a priority. The Kabinett designation falls into the Qualitätswein category, with parameters set for specific sugar and alcohol levels.

Tafelwein is almost like it sounds—"table wine"—and is the Austrian version of the French Vins de Pays and Italy's Vino da Tavola, typically representing the lower price points for everyday wines. Landwein, or "land wine," also falls under the Tafelwein category. Landwein is considered somewhat better on the quality scale because it's made from allowable grape varieties only and must label its region of origin.

LOWER AUSTRIA

The Region and the Wine

By far the largest and most significant wine-growing region in Austria, Lower Austria—named for its location on the Danube River, not the geographical position within Austria—is made of four key subregions: Donauland, Kamptal, Kremstal, and Wachau. Each of these areas are featured on the wine's label, as opposed to the more general "Lower Austria." White wine is the star here, with the food-friendly grape duos of Grüner Veltliner and Riesling playing starring roles and Müller-Thurgau, Chardonnay, and Sauvignon Blanc playing vineyard backups.

Lower Austria region

Grüner Veltliner (a.k.a. GV, GrüVe, or just Grüner) is Austria's primary white wine grape and accounts for about a third of the nation's vineyard plantings. Many other grapes are grown locally, but this is the grape that garners worldwide export attention. Grüner Veltliner produces a refreshingly crisp, citrus-heavy wine with a hint of herbal and often white pepper nuances on the finish, with frequent doses of earthy minerality and awesome food-craving acidity. This is one of the few grapes that maintains a certain affinity for vegetables, even embracing the hard-to-pair asparagus and artichoke.

BURGENLAND, AUSTRIA

Burgenland region

The Region and the Wine

Situated along the eastern Austrian border alongside Hungary, Burgenland is home to Austria's most notable red wines and has earned a worldwide reputation for its rich dessert wines. Major subregions in Burgenland include Neusiedlersee (named after the large, shallow, influential regional Lake Neusiedl) and Mittelburgenland, both of which grace wine labels. The dominant red wine grapes are Blaufränkisch (or Lemberger), which can exhibit bold fruit character with ripe, raspberry flavors, a spicy component, velvety tannins, and a plush mouthfeel; and Zweigelt, a lighter-styled red with more subtle tannins. St. Laurent and Spätburgunder (Pinot Noir) account for the region's other key red wine grapes. The Burgenland's decadent, full-bodied, full-flavored dessert wines are made predominately from white grapes that have been either completely or partially affected by botrytis, a "helpful" fungus that concentrates the grape's innate sugar content.

The relatively warmer Continental growing conditions in the Burgenland allow for greater grape ripening than in many other Austrian zone, which is why red wine grapes are grown with greater concentration in the Burgenland. Local lakes, with Lake Neusiedl the most prominent, contribute humidity that helps foster the growth of botrytis on the Furmint, Chardonnay, Traminer, Weißburgunder, Welschriesling, and Riesling grapes that remain on the vine past the traditional harvest. These shriveled, well-concentrated grapes are used to make the famous Ausbruch, Beerenauslese, and Trockenbeerenauslese sweet wines.

Selecting and Pairing

Austrian cuisine consists of flavorful, hearty fare. From schnitzel to strudel and goulash to spätzle (the local version of macaroni and cheese), much of Austria's rich food is based on fresh, local, seasonal sources. Fortunately, the regional wines are very well suited to pair with many of these dishes. Austria's white wines, Grüner Veltliner and Riesling, are among the most food-friendly wines around.

MEAT

The spicy, berry character of Blaufränkisch is a natural match for meaty dishes. Whether it's the local goulash with beef, noodles, and tomatoes or schnitzel, bratwursts, and burgers, the ripe flavors and easygoing tannins provide lots of food-friendly diversity. Understated and lighter-tannin red Zweigelt pairs with pork cuts and roasted lamb extraordinarily well. Austrian Spätburgunder (Pinot Noir) is an up-and-coming grape and provides a delicious pairing with many cuts of beef. Locally, white wines are often paired with various sausages, and Grüner Veltliner showcases smoky, spice-laden meats particularly well.

POULTRY

Roasted poultry is often served alongside a variety of vegetables. Grüner Veltliner shines amid the finicky chemistry of asparagus, artichokes, broccoli, and the like. Capable of bringing out the best in the food and the wine, Grüner's accommodating character complements roasted duck or turkey, fried chicken, and chicken salad. Sweet Ausbruch is a top pick with the decadent flavors of foie gras.

SEAFOOD

Lower in both alcohol and tannin, Zweigelt makes an easy partner for fresh fish. For white wine and seafood, there's nothing better than the aromatic, high acidity profile of Grüner Veltliner, whether it's with breaded abalone, shellfish, fried fish, or raw fish in sushi with spicy Asian themes.

CHEESE

Spätburgunder is a perfect partner for cheese. Try it with Brie, Camembert, feta, Gruyère, Swiss, Monterey Jack, or Muenster. Grüner Veltliner is a natural alongside pungent, zesty herbed goat cheese. For the stronger aromas and tanginess of blue-veined cheese, consider Austria's Ausbruch dessert wine.

DESSERT

The regional Eiswein (ice wine), with its delicious balance of sweetness and acidity, makes a remarkable partner for creamy custards, fresh berry tarts, and cakes. The richer, sweeter Beerenauslese and Trockenbeerenauslese are natural pairs for apple strudel, crème brûlée, and sweet almond pastries.

PRODUCERS TO TRY

Fritsch, Gobelsburger, Hirsch, Hirtzberger, Huber, Kracher, Laurenz V., Nigl, Nikolaihof, Nittnaus, Pichler, Prager, Weingärtner, Zantho

CZECH REPUBLIC

GERMANY

SLOVAKIA

UKRAINE

AUSTRIA

HUNGARY

ROMANIA

BLACK SEA

ITALY

SERBIA

BULGARIA

TURKEY

GREECE

Barcelona

MEDITERRANEAN
SEA

WINES OF THE REST OF EUROPE

Romania, Hungary, and Greece might seem like an odd team of wine players to discuss together, given their widely varying cultural, political, and geographical differences. Their common wine denominator ties all three nations' wine-making history back to at least the Roman Empire (and Greece considerably before), and together they make up about 7 percent of Europe's total wine production.

Although individual wine volume weighs in on a much smaller scale than that of familiar favorites like France and Italy, these three countries have been making serious strides in their wine's quality and market availability. In fact, the last two decades have seen a remarkable increase in overall quality for Romania and Hungary as communism lost its grip, leaving much of the generic, mass-produced wine in the past while capital investments from neighboring European countries provided for the much-needed modernization of the wineries and vineyard upgrades.

Greek wines also experienced a significant leap in quality levels with their inclusion into the European Union wine-making community in the early 1980s.

GREECE, HUNGARY, AND ROMANIA

Greece, Hungary, and Romania all lean heavily on a broad spectrum of indigenous grapes, with significant ongoing experimentation taking place among key international varieties. Most of the wines produced in these countries are consumed within their own borders, yet limited exports are gaining a foothold, especially in neighboring European countries. Some are beginning to see more shelf space abroad, too.

Greece

Greece hosts several hundred varieties of native grapes within its incredibly varied landscape. Islands, mountains, and seas combine in a wild mix of terrain and terroir to produce a rare collection of varying microclimates that can lean toward Mediterranean or Continental, depending on specific vineyard locations.

The vast majority of the country's wine offerings are white, with three major grape varieties: Assyrtiko, which produces a mineral-driven, crisp, dry white wine most often found in Santorini that's perfect for the locally abundant seafood; Moschofilero, the incredibly aromatic white wine grape that brings a spicy citrus character to the glass and makes a delicious partner with the fresh, exotic flavors of sushi and Asian or Middle Eastern cuisine; and Agiorgitiko, the most popular red grape variety that creates delicious everyday red wines spotlighting cherry and spice with typically well-managed acidity and delicate, velvety tannins, perfect for pairing with lamb-themed stews, gyros, or pitas with Greek salad accompaniments.

Greece also has a sweet dessert wine delight, Vinsanto, which boasts a sweet nectar character and is made traditionally from well-ripened grapes that are harvested and then dried outdoors for about 2 weeks prior to fermentation and oak-barrel aging.

Greece ■ Hungary ■ Romania

Hungary

Most famous for its exquisite and expensive dessert wine, Tokaji Azsú (which we look at later), Hungary also produces a wide range of red and white table wines in both dry and off-dry styles and continues to be a trailblazer for Eastern Europe's up-and-coming wine scene.

Furmint, the main grape in Tokaji Azsú, is also put to work in a variety of exotic, fruit-driven, dry white wines with crisp acidity and an engaging, food-friendly nature. Hárslevelű is another one of the three Hungarian white wine grapes used to make Tokaji Azsú, but it also enjoys life on its own in a dry, highly aromatic white varietal wine. Kadarka, a popular red wine grape, tends to go subtle on the tannin and high on the spice, with plenty of ripe, red fruit in between, perfect for pairing with beef stews, earthy mushrooms, and a bit of black pepper and paprika spice.

Key international varieties known and grown in Hungary include Pinot Noir, Merlot, Cabernet Sauvignon, Chardonnay, and Sauvignon Blanc.

Romania

The Carpathian Mountains, the Danube River, and the Black Sea all influence Romanian vineyards and culminate to produce a largely Continental climate zone, ideal for growing a tremendous variety of grapes.

Fetească Neagră ("black maiden"), typically a medium-bodied red that can be made in a dry or off-dry style with dark fruit accents, is perfect for pairing with meat and mushroom dishes, smoked game, and even aged cheeses. Its sister grape, Fetească Albă ("white maiden") is a native white grape that makes a light- to medium-bodied wine that leans toward green apple and some citrus and partners well with fresh seafood.

Cabernet Sauvignon, Merlot, Pinot Noir, Syrah, Pinot Gris, Sauvignon Blanc, Chardonnay, and Riesling round out some of Romania's top international varieties.

WINES OF NORTH AMERICA

The United States and Canada make up the North American wine scene, with California, Washington, Oregon, New York, and Virginia representing the top wine-producers in the United States, and Canada's Ontario and British Columbia dominating domestic production with still, sparkling, and dessert wines.

Every state in the union is now making wine (although not all are growing grapes) in some form or fashion, but California reigns as the top North American wine producer. According to the Wine Institute, California produces 90 percent of all the wines made in the United States and represents the fourth-largest wine-producing region in the world, after the big leagues of France, Italy, and Spain.

There's plenty to be excited about in terms of New World wine production taking place in North America. Growing regulations tend to focus on defining appellations and not on heavy-handed dictates for specific vineyard management or wine-making protocol. Freedom to experiment with a variety of grapes, barrels, and wine-making interventions while taking cues from traditional growing regions has allowed many of the winemakers in North America to propel their wines to the top of their given tiers. In fact, the vast majority of Americans pick domestic producers over imports, perhaps due to the easy familiarity of grape names on labels, a general knowledge of certain American growing appellations, or the full-fruit appeal of many North American wines.

With so many places planting grapes, and ever-increasing enthusiasm for wine itself, the culture of wine in North America is brighter than ever.

AMERICAN VITICULTURAL AREAS

American Viticultural Areas, or AVAs, are specific regions of wine production in the United States. Somewhat similar to Europe's appellations of origin, AVAs are federally recognized, uniquely delineated areas for growing wine grapes that are distinguished by various measures of terroir, with special consideration given to topography, geography, and climate.

The size and scope of an AVA can range from tiny parcels of land to enormous spans of earth like that found in Washington's Columbia Valley—which comprises a third of the state! Within larger AVAs, smaller, more descript subregional AVAs may be found.

So if a winery would like to showcase fruit from a small slice of the Napa Valley, for example, it might feature grapes grown only from the Stags Leap District (and labeled as such). However, if it would like to source grapes from the Napa Valley at large, the label would reflect "Napa Valley." Still, perhaps the winery would like to make an easygoing, everyday table wine and doesn't want to pay the premiums for prestigious Napa Valley grapes. In this case, it might branch out even further to larger AVAs and simply note "California" AVA on the label (although most labels don't record "AVA," like the French AOC or Italian DOC).

American AVAs are strictly geographical in nature. They don't attempt to regulate which grapes may be grown where or set parameters for vineyard yields, dictate aging protocol, or attempt to establish any other vineyard management or wine-making practices. In true New World fashion, America's various AVA label designations take a backseat to the top priority: the starring role of the leading grape varietal.

U.S. Wine Laws

Wine laws in the United States are similar in style to the AVA designation, in that they set some structure but don't dictate tremendous detail.

All wines labeled by grape varietal must have a minimum of at least 75 percent of that grape type in the bottle, leaving a bit of wiggle room for blending in complementary (or at times cheaper) grapes. If a specific AVA is designated, at least 85 percent of the grapes must be sourced from within those geographical parameters.

As far as vintage years are concerned, 95 percent of the wine in the bottle must be from the featured year.

American wine laws are refreshingly straightforward compared to many European wine-growing nations, allowing for vineyard managers and winemakers to determine best practices for their particular bottling efforts.

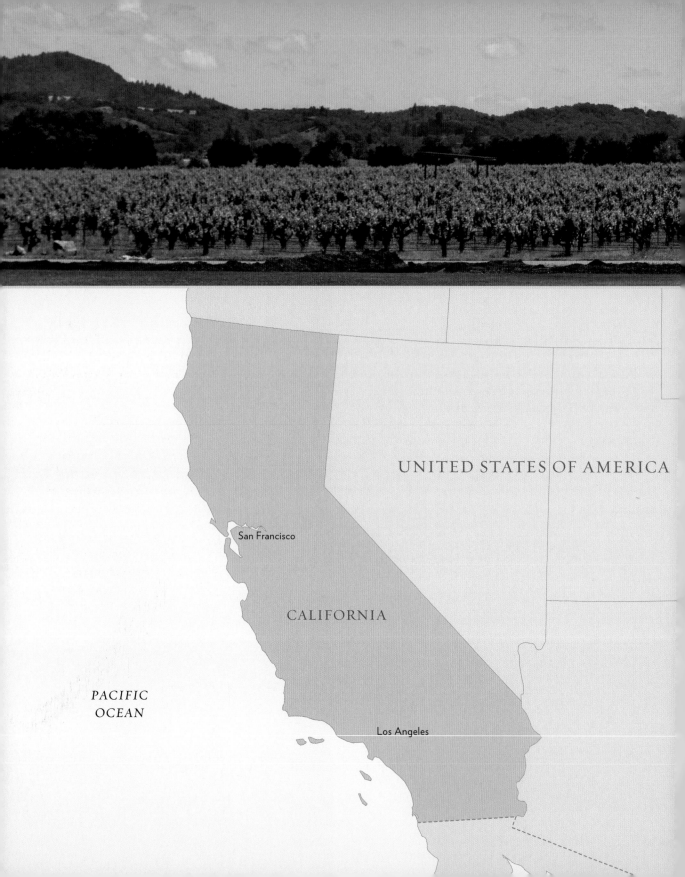

PACIFIC
OCEAN

CALIFORNIA

San Francisco

Los Angeles

UNITED STATES OF AMERICA

WINES OF CALIFORNIA

California's wine history is marked by variety, versatility, and innovation. The world's fourth-largest wine-producing region (after France, Italy, and Spain), California accounts for 90 percent of the United States' total wine production, or more than 200 million cases of wine annually, with exports numbering right around 50 million. Close to 3,500 wineries call California home. The Pacific Ocean plays an enormous role in the diverse range of microclimates found in the region's 115 American Viticultural Areas (AVAs), as do the sun-soaked sections of the Central Valley.

California is the epitome of a New World wine-growing region, embracing technology yet tipping its hat toward the art, science, and tradition of the Old World. California's wine history dates to the late 1700s and the Spanish missionaries, who cultivated vines primarily for sacramental and ceremonial purposes. The mid-1800s brought the Gold Rush, and vineyards were hard-pressed to meet the demands of ambitious gold diggers. Phylloxera, a root-eating louse, devastated California's vineyards (and the rest of Europe's) by the late 1800s. Fast forward to the 1920s, and Prohibition brought all serious vineyard endeavors to a screeching halt until its repeal in 1933.

It wasn't until the early 1970s that California's wine renaissance arrived, thanks to Robert Mondavi's persistent passion and unfaltering vision that California could make world-class wine to rival the best of Europe. His intuition proved true in the legendary 1976 Paris Tasting (also called the Judgment of Paris), where Napa Valley's 1973 Chateau Montelena Chardonnay and Stag's Leap Wine Cellar's 1973 Cabernet Sauvignon rocked the world by winning a blind-tasting (by French judges) against the French wine giants, Burgundy and Bordeaux. This win thrust California onto the international wine stage, and it's never looked back.

Today, innovation, research, and experimentation abound, as do more than 100 different grape varieties grown across the state. In keeping with its initial story of success, California Cabernet Sauvignon and Chardonnay are still the stars, especially in Napa Valley, Sonoma County, and the Central Coast, where the great majority of the state's wine grapes are grown.

Napa Valley, California

The Region and the Wine

Napa Valley, the star of California's wine scene, is an hour's drive north of San Francisco. Situated between the Mayacamas Mountains on the west and the Vaca Mountains on the east, the Napa Valley AVA is a mere 30 miles (48.25 kilometers) from top to bottom and 5 miles (8 kilometers) across. With sunny, warm days and cool nights, the valley is a winemaker's dream, consistently producing world-class wines from more than 400 wineries. Thirteen smaller AVAs are tucked within Napa Valley, the most important being Stags Leap District; Rutherford; Oakville; Spring Mountain; Howell Mountain; Mount Veeder; and a significant slice of the foggy, cooler climate of Carneros.

Windsor

PACIFIC OCEAN

■ Napa County

Despite representing less than 5 percent of California's total wine production, Napa Valley remains the United States' premier wine-growing region. Cabernet Sauvignon is Napa Valley's signature red wine grape, with its clear expressions of dark berry fruit, vanilla, and toasted oak, along with well-honed tannin structure (making aging an option for a decent percentage of Napa's top stock). Merlot, Zinfandel, and Pinot Noir (especially from cooler Carneros) round out the region's most prominent red wine grapes. Chardonnay is the most widely planted white wine grape. Tart green apple; lively citrus notes; and a rich, well-oaked, buttery character comprise classic Napa Valley Chardonnay. A close second, Sauvignon Blanc offers a lighter-styled balance to Chardonnay's weight and more complicated character.

Selecting and Pairing

Napa Valley places a priority on fresh, local, and seasonal cuisine. The epicenter of agricultural prowess for the entire United States, California has a habit of delivering ultra-ripe fruits and vegetables, fresh seafood, artisanal olive oils and cheese spreads, along with an abundance of local poultry and beef.

MEAT

Whether it's a flavorful peppercorn steak or a grilled New York strip, Napa's big, bold reds pair well with the robust flavors of regional beef dishes. For the peppery portion, pick a Napa Zinfandel; for a well-marbled rib eye, opt for the tighter tannin of Napa's Cabernet Sauvignon.

CHEESE

Fresh, artisanal cheese abounds around Napa Valley. With goat's, sheep's, and cow's milk cheese options, regional wines find plenty of friends. Consider the ripe cherry nuances and malleable tannin of Merlot to complement the flavor profiles of smoked Gouda, Camembert, or Jarlsberg cheeses.

POULTRY

As a partner for some of the region's best free-range poultry, a cool-climate Pinot Noir from Napa's Carneros region suits. Its sleek, lighter-bodied style and upfront berry makes the match with a variety of chicken dishes, turkey cuts, and even roasted game birds. Chardonnay marries well if there's a bit of butter in the dish; if herbs prevail, opt for the fresh, earthy, sometimes grassy nature of Napa's Sauvignon Blanc.

DESSERT

Bright berry desserts, peach cobbler topped with vanilla ice cream, fresh marscapone inspirations, or fruit sorbet—whatever your craving, a late harvest Riesling or dessert Muscat partners sweet with sweet in delicious harmony.

SEAFOOD

Napa's easy access to the coast provides the full spectrum of fresh fish and shellfish. However, the local Dungeness crab rocks the restaurant scene with its sweeter flavors and generous portions of tender meat often drizzled in butter—a perfect pairing for the rich, creamy nature of Napa Chardonnay.

PRODUCERS TO TRY

Beringer, Beaulieu Vineyard, CADE, Cakebread Cellars, Caymus, Chateau Montelena, Clos Du Val, Domaine Carneros, Duckhorn, Étude, Far Niente, Frog's Leap, Grgich Hills, Hall, Hess Collection, Joseph Phelps, Mount Veeder, Pine Ridge, Robert Mondavi, Round Pond Estate, Rutherford Ranch, Schramsberg, Shafer Vineyards, Silver Oak Cellars, Smith Madrone, Spotteswoode, St. Supéry

SONOMA, CALIFORNIA

The Region and the Wine

Considerably larger and remarkably laid back compared to its ritzy Napa neighbor, Sonoma County's maritime influence often offers cool, foggy mornings, especially closer to the coastline, and warm summer days. A unique mix of forests, fruit orchards, and vineyards dot Sonoma's landscape with the Pacific Ocean to the west and the Mayacamas Mountains separating Sonoma from the Napa Valley on the east. With 15 AVAs encompassing hillside vineyards to expansive vineyards along the valley floor, Sonoma's distinct regional microclimates give rise to dramatic topography and great soil diversity, encouraging the successful growing of a broad range of grapes.

■ Sonoma County

Although Sonoma's twice the size of Napa Valley, it typically only makes half the volume of wine. The warmer Alexander Valley AVA concentrates on cultivating stellar Cabernet Sauvignon, while cooler coastal regions like the Russian River Valley tend toward Burgundy's best grapes, Chardonnay and Pinot Noir, producing highly versatile wines boasting both elegance and endurance, with the bright-berried, silky tannin structures of regional Pinot Noir and Chardonnay carrying classic acidity and deep minerality.

Selecting and Pairing

Farmers' markets emphasize local and keep Sonoma's regional cuisine fresh and vibrant while inspiration is also drawn from international fare. Seafood, poultry, and beef are easy cooking commodities given the proximity to the Pacific and abundance of regional farms.

MEAT

The full flavors of various beef cuts, sausages, wild game, bison, and pork tenderloin marry extremely well with the rich structure, well-managed tannin, and plush fruit of Sonoma County Cabernet Sauvignon or the spicy, peppery character of regional Zinfandel.

CHEESE

Pinot Noir, with its silky tannin and pronounced red fruit partners well with the soft, creamy lines of Brie or Camembert. In the whites, Chardonnay offers a fairly full-bodied profile to handle Brie's rich textures.

POULTRY

The rich, round, creamy flavors of Sonoma County Chardonnay shine next to roasted chicken, chicken Alfredo, and even Parmesan-crusted chicken. Heartier duck and goose also appreciate the impressive versatility of the full flavors and lively acidity in the regional Chardonnay.

DESSERT

Sonoma's late harvest wines make perfect pairings with an assortment of sweets. Consider a late harvest Riesling for apple pie, crème brûlée, angel food cake topped with berries and cream, or simple shortcake cookies with almond accents.

SEAFOOD

The citrus flavors, creamy textures, and buttery nuances of Sonoma Chardonnay play well with crab cakes, macadamia-crusted halibut, grilled salmon steaks, and lobster bisque. The dry, crisp character of Sauvignon Blanc brings out the best in shellfish and lighter fillets of fish.

PRODUCERS TO TRY

Benziger, Chalk Hill, Chateau Souverain, Chateau St. Jean, Ferrari-Carano, Freestone Vineyards, Gary Farrell, Geyser Peak, Iron Horse, J Vineyards, Kendall-Jackson, Kunde Family Estate, La Crema, Louis M. Martini, MacMurray Ranch, Paul Hobbs, Quivira, Ravenswood, Rodney Strong, Sebastiani, Simi

California's Central Coast

The Region and the Wine

From San Francisco County down to Santa Barbara County, California's Central Coast spans nearly 300 miles (483 kilometers) long and an average of 25 miles (40 kilometers wide, boasts more than 90,000 acres (36,421.75 hectares) under vine, and is home to more than 350 wineries. Tracing the Pacific coastline, with its maritime reach extending farther into some regions than others, and buffered by the hot temperatures of the Central Valley, the Central Coast AVA grabs the best of both worlds, catching warm days, cool nights, and early morning fog closer to the coast. The region can be divided into the northern counties, central counties, and southern counties; the San Francisco Bay Area dominates the north, Monterey and San Benito comprise the central counties, and San Luis Obispo and Santa Barbara round out the southern sphere.

■ Central Coast region

In the north, Chardonnay, Cabernet Sauvignon, and Merlot grape varieties reign. The cooler climes of the central coast's Monterey County shine with Chardonnay, Pinot Noir, and Merlot. In San Luis Obispo County, Paso Robles is the rising star distinguished by its many remarkable Rhône-style blends, although Cabernet Sauvignon and Merlot are the most widely planted grapes. Santa Barbara County takes advantage of its cool, maritime climate, thanks to the east-west orientation of its coastal mountain range, and enables Chardonnay and Pinot Noir grapes to enjoy longer hang-time on the vine, maximizing acidity and flavor maturity.

Selecting and Pairing

California's cuisine finds terrific footholds in the seasonally inspired farm-to-table movement, giving growers, chefs, and localvores some serious bragging rights. With easy access to The Abalone Farm in Cayucos and fresh Morro Bay oysters and Dungeness crab, the Central Coast bursts with tremendous seafood options.

MEAT

Whether it's wild game, beef Provençal, pork tenderloin, roasted rack of lamb, or braised baby back ribs, Rhône-style blends from Paso Robles can handle it all with bright red fruit, easygoing elegance, and well-honed structure. The dense, dark fruit; modest tannin; and bold profiles of the regional Cabernet Sauvignon also partners well with a variety of meat themes.

CHEESE

Central Coast Cabernet Sauvignon befriends a variety of well-aged, hard-profiled cheeses like cheddar, Colby, Parmesan, and the soft but pungent Danish Blue, as the wine's well-worn tannin structure softens the firm edge of full-flavored cheese spreads. Chardonnay's creamy textures and ripe fruit nuances make soft-styled cheeses like Brie and Camembert a top pairing pick.

DESSERT

Late harvest and fortified wines from California's Central Coast turn fruit desserts into a delicious duo. The heavier textures and sweeter styles of both late harvest and fortified wines complement the similar weights and flavors of peach cobbler, a medley of berries topped with cream, and custard-style delights.

POULTRY

Taking the cornucopia of flavors in an Asian stir-fry often mixed with chicken and varied seasonings, the Rhône-style white blends from Paso Robles or Santa Barbara's Viognier boast vivid fruit flavors and delicious diversity. Regional Chardonnay carries poultry favorites with rich, creamy style and persistent palate grace.

SEAFOOD

The Central Coast is home to an abundance of ultra-fresh seafood, all perfect for pairing with a variety of lively white wine blends whose ample acidity cuts through a variety of tastes, textures, and briny character. Heady aromas, stunning tropical fruit, and a rich and creamy character make regional Viognier an excellent choice for shellfish.

PRODUCERS TO TRY

Au Bon Climat, Bridlewood Estate, Curtis Winery, Eberle, Epiphany Cellars, Fess Parker, Firestone Vineyards, J. Lohr, JUSTIN Vineyards, Kenneth Volk, L'Aventure, Saxum, Stolpman Vineyards, Tablas Creek, Treana, Vina Robles, Zaca Mesa

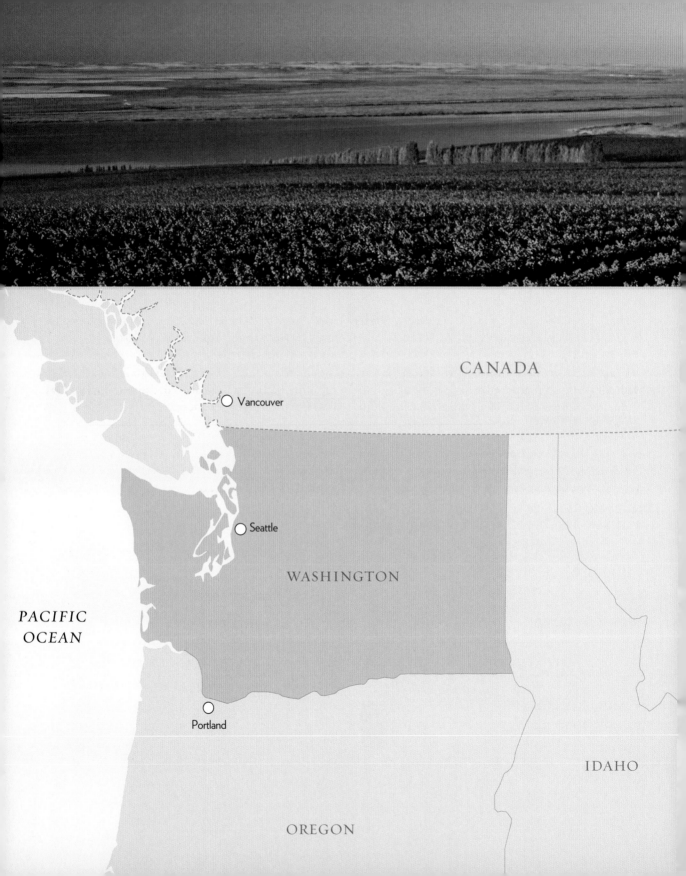

PACIFIC
OCEAN

CANADA

○ Vancouver

○ Seattle

WASHINGTON

○
Portland

IDAHO

OREGON

WINES OF WASHINGTON

The number-two wine producer in the United States, Washington has close to 40,000 acres (16,187.5 hectares) under vine and produces 12 million cases of wine annually. The last two decades have seen thrilling surges in both quality and quantity—more than 30 unique grapes—grown in the state, with special emphasis on Merlot, Cabernet Sauvignon, Syrah, Chardonnay, and Riesling. Washington's red wines bring remarkable concentration, supple structures, and ongoing fruit essence. Its whites burst with delicious fruit character and bracing acidity, taking full advantage of the warm, sunny days and cool, crisp evenings found east of the Cascade Mountains. Washington's wines currently run with a 55–45 split of white wines to red.

Washington boasts 112 American Viticultural Areas (AVAs), and the Columbia Valley is the largest, engulfing 8 smaller sub-AVAs. Most of the state's 750 wineries and 350+ grape growers are on the eastern side of the Cascades, which act as a rain shield, forcing clouds to dump moisture on the western slope and leaving the eastern plains dry and desertlike. The dramatic desert landscapes are highlighted by the lush irrigated vines within. Bordered by nothing but dirt, the stunningly stark nature of eastern Washington's hillside scene seems to embrace the green gift of life the orchards, vineyards, and concentrated valleys of produce bring.

Two of the best Old World producers maintain partnerships with the Washington powerhouse Chateau Ste. Michelle. The Mosel's Ernst Loosen (Dr. Loosen on labels) has joined forces with Chateau Ste. Michelle to create the remarkable Eroica Riesling. This well-priced Riesling highlights the grape's energetic appeal; off-dry, semi-sweet style; and well-managed acidity, showing the super synergy created by two of the world's most formidable Riesling advocates. Chateau Ste. Michelle also teamed up with Tuscany's Antinori to create Col Solare, a Bordeaux-style blend of Cabernet Sauvignon (usually around 75 percent) with Merlot and Cabernet Franc.

Washington's wine story is just revving up; the state's only been seriously pursuing wine production for the last couple decades. New wineries are routinely springing up to test their vinification skills on the worldwide market. With so many delicious, well-concentrated wine options at reasonable price points, Washington's wines have a lot to celebrate.

COLUMBIA VALLEY, WASHINGTON

The Region and the Wine

The Columbia Valley is Washington's largest AVA. Covering more than 11 million acres (more than 4 million hectares), and responsible for 99 percent of the region's wine grape production, the Columbia Valley is comprised of 8 sub-AVAs: Walla Walla, Yakima Valley, Red Mountain, Horse Heaven Hills, Rattlesnake Hills, Snipes Mountain, Red Mountain, and Wahluke Slope. Enveloping a third of the state's land, the vineyards see a remarkable range of microclimates, with most falling in the arid desert climate zone east of the Cascades and irrigated by the Columbia, Yakima, and Snake Rivers. The generally warm, dry vineyards typically see less than 10 inches (25.5 centimeters) of rain per year. The soil is mostly sand, silt, and granite, and many of the region's vines grow on south-facing slopes to catch more sunlight.

■ Columbia Valley

More than 30 grape varieties are grown in Washington, and Riesling, Chardonnay, and Sauvignon Blanc are the stars. These whites boast bracing acidity, with focused fruit character and a delicious range of styles. Merlot, Cabernet Sauvignon, and Syrah are the state's top reds, with well-built yet supple tannins, powerful profiles, lush textures, and concentrated and robust berry aromas and flavors. Washington's mix of warm, sunshine-filled days and cool, crisp nights keep grape acidity in top shape.

Selecting and Pairing

The brilliant acidity that marks many of Washington's top wines brings exceptional food-pairing capabilities to a variety of flavors, styles, and international influences. An important agricultural region, Washington bases its regional cuisine on fresh, ripe, and ready food sources, not to mention Seattle's seafood.

MEAT

The regional Merlot, Cabernet Sauvignon, and Syrah were made for meat-themed entrées. Supple yet potent tannins handle the fat and protein found in beef, sausage, and full-flavored game with elegance and complete palate appeal. Merlot tops the charts for lamb chops.

POULTRY

The concentrated plum and berry tones of Washington Merlot make a remarkable match with a variety of baked chicken and pasta dishes, grilled duck, and roasted turkey recipes. Try Syrah with Peking duck. Earthy herbal accents on poultry call for the equally earthy nature of Sauvignon Blanc.

SEAFOOD

Washington's Riesling works well with just about every one of Seattle's choice seafood selections. For butter-laden fare, opt for a round, fuller-bodied Chardonnay. With grilled shellfish or seafood salads, reach for a slightly grassy, citrusy Sauvignon Blanc.

CHEESE

The spicy, concentrated fruit of a Washington Syrah supports the full flavors of a rich, well-aged Parmesan or Gouda. Riesling or Chardonnay offers both ends of the style spectrum for partnering with the soft textures and subtle nutty nuances of a wheel of Brie.

DESSERT

With so many late harvest dessert wines available in and around the Columbia Valley, many based on the charms of the Riesling grape, the dessert pairing options are almost endless. Try pairing the sweetest side of Riesling with the abundant, local apple–based desserts, almond shortcake, peaches and cream, or bread pudding—or enjoy solo as dessert itself.

PRODUCERS TO TRY

Betz, Canoe Ridge, Chateau Ste. Michelle, Chester-Kidder, Col Solare, Columbia Crest, Covey Run, Gordon Brothers, Hogue Cellars, L'Ecole N° 41, Leonetti Cellar, Mercer Estates, Pedestal, Northstar, Quilceda Creek, Seven Hills, Woodward Canyon

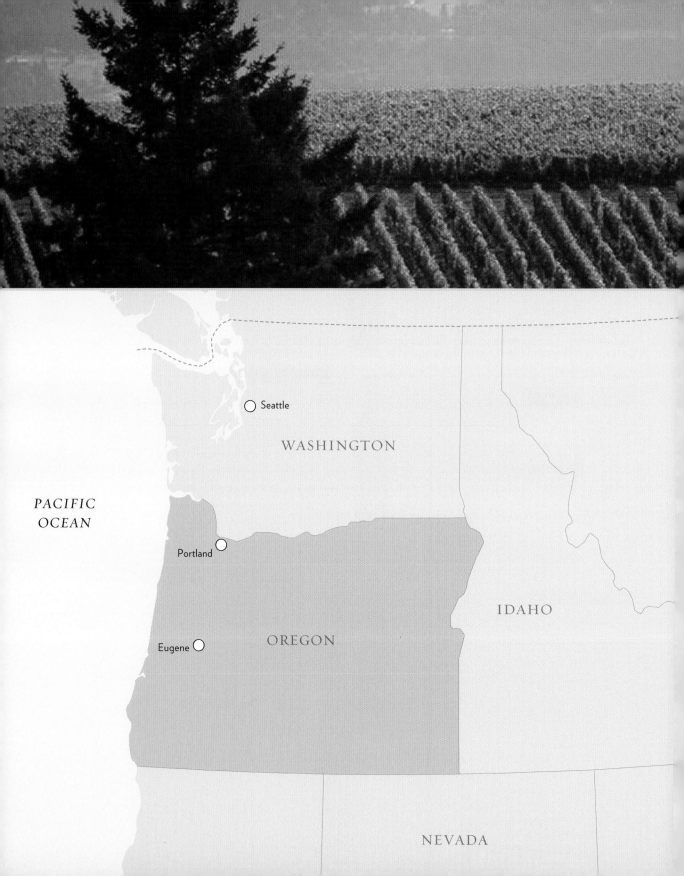

PACIFIC
OCEAN

Seattle

WASHINGTON

Portland

OREGON

Eugene

IDAHO

NEVADA

WINES OF OREGON

Oregon's wine endeavors began in the 1970s, when cutting-edge winemakers became convinced that cool-climate grapes could not only survive but also thrive in the Pacific Northwest. With a keen focus on French vines, Pinot Noir, Pinot Gris (same grape as Italy's Pinot Grigio), and Chardonnay were imported for their ability to grow well in the cool climates of both Alsace and Burgundy. Decades later, these grapes sit atop the Oregon's grape-growing hierarchy.

Oregon is the fourth-largest wine-producing state, with more than 400 wineries calling it home. The vast majority of the state's wineries are small, boutique, and family owned.

Oregon contains 16 recognized American Viticultural Areas (AVAs), with the most production coming from the Willamette Valley, named after the major river that flows from Portland south to Eugene. Other key appellations—the Umpqua Valley, Rogue Valley, and Applegate Valley AVAs—represent the next chapter in Oregon's ongoing viticultural story.

Pinot Noir is queen of the state, representing more than 50 percent of the cultivated vines. Oregon Pinot Noir reigns with earthy personas and vivid, focused red fruit character often accompanied by ambitious acidity, silky tannins, and an enduring elegance. Pinot Gris and Chardonnay are the key players in the region's white wine circle, and both are capable of producing wines of character and complexity, with the fresh factor of Pinot Gris bringing a seafood-friendly compatibility—and often a very reasonable price tag.

Oregon's climate can be tricky, thanks to plenty of rain, moderate sunshine, and the marginal weather patterns that leave the vines and vintners guessing and gambling as to what will come next. However, the gambles pay off in good vintages. The wine rewards are great, giving the region palate-pleasing vino that exudes elegance, complexity, and ethereal character.

OREGON'S WILLAMETTE VALLEY

The Region and the Wine

Willamette Valley

With cool climates, well-drained soils, hillside vines, and the protection from mountains flanking both sides, Oregon's Willamette Valley escapes the brunt of the Pacific coast storm cycles and enjoys a unique grape-growing environment. The fussy nature of the Pinot Noir grape prevalent here makes plenty of demands, including a cool season for ripening and maturing. Creating the elegant, earthy wines the Willamette Valley is known for, cooler temperatures, site-specific vineyard orientations, vine age, unique vine clones, and winemaker influences all impact Oregon's vines and wines.

Burgundy is Pinot Noir's first home, but Oregon isn't far behind. Here, Pinot Noir tends to take on a pronounced fruit-focused character. The earthy forest floor and mushroom aromas remain, but they often take a backseat to the brighter berry and cocoa profiles that emerge from Willamette Valley fruit. Chardonnay keeps a tight rein on its weight in Oregon, with racy, lean palate profiles. Pinot Gris enjoys rich textures, lovely aromatics, and persistent fruit, while maintaining astonishing acidity and food pairing performance.

Selecting and Pairing

Fresh, local food shines bright throughout the Willamette Valley, with plenty of seafood options gracing menus and regional farms providing abundant seasonal produce. Thankfully, Oregon's food-friendly trio of Pinot Noir, Pinot Gris, and Chardonnay are more than capable of bringing out the best in local fare.

MEAT

Lamb chops with rosemary, beef with mushrooms, roasted game in soups and stews, smoked meats with fresh herbs, and ham or prosciutto are all perfectly at home with the earth, finesse, and fruit of Oregon's provocative Pinot Noir.

CHEESE

Pinot Noir is a long-time friend of most cheeses. Brie, Swiss, Gruyère, and some local goat's cheese options work well with the understated tannins and earthy profile of Oregon Pinot Noir.

POULTRY

Roasted, grilled, baked, stuffed, and sautéed chicken dishes are perfect for Oregon's Pinot Gris. The svelte local Chardonnay handles a fairly broad spectrum of poultry recipes with ease, especially those starring butter, cream, and cheese.

DESSERT

Oregon's rich dessert wines and Port offerings complement a range of sweets. With light and fruit-inspired meringues, tarts, pies, shortbread cookies, and assorted cheesecake, Oregon's dessert wines more than shine. For dense, dark chocolate specialties, opt for the power and profile of a Willamette Valley Port.

SEAFOOD

Pinot Noir partners extraordinarily well with fresh, grilled salmon or tuna, marrying the oily content of the fish with the grease-cutting acidity of the wine. The regional Pinot Gris is a delicious choice for regional seafood favorites, from Dungeness crab chowder to shrimp cocktail, clams, scallops, and steamed mussels.

PRODUCERS TO TRY

Adelsheim, Archery Summit, Argyle, Bethel Heights, Chehalem, Domaine Drouhin, Eyrie, King Estate, Ponzi, Rex Hill

CANADA

NEW YORK

ATLANTIC
OCEAN

UNITED STATES OF AMERICA

WINES OF NEW YORK

New York is the third-largest wine-producing state in the United States, boasting more than 200 wineries. It's also reputed to be one of the oldest commercial wine-growing areas in the country. New York is home to nine nationally recognized American Viticultural Areas (AVAs). The Finger Lakes, Hudson Valley, and Long Island constitute the three most important AVAs today.

Both Native American (Concord grape juice) and European grape varieties (Riesling, Chardonnay, Syrah, and other international varieties) stake considerable claims in New York's vineyards, as do several significant French American hybrids (crossed for cold-resistance, like Vidal).

The European grape varieties have a relatively recent history, thanks to the research, work, and optimism of Ukrainian viticulturist Dr. Konstantin Frank in the mid-1960s, followed by his son's continued enthusiasm for creating world-class wines in New York's Finger Lakes region in the mid-1980s. Their combined passion for growing Old World grapes, along with their practical know-how for managing cold-climate vineyards, changed the course of New York's wine industry. The tangible results of Dr. Frank's viticultural endeavors have laid the groundwork for significant, successful wine production and inspired a generation of grape growers to pursue European varieties within their own vineyards.

NEW YORK'S FINGER LAKES

The Region and the Wine

Western New York's Finger Lakes produce the vast majority of New York's wine, harvesting 40,000 tons (36,287.5 metric tons) of grapes a year from 10,000 acres (4,047 hectares) of vine. With just over 100 wineries, the region focuses on a range of wines, including Riesling, Gewürztraminer, and Pinot Noir, along with sparkling wines and remarkable ice wines. New York Riesling grows incredibly well in the cool-climate zones of the Finger Lakes, often producing wines of extreme character, showcasing zesty, citrus fruit; potent aromatics; and a dynamic balance of sugar and acidity.

The cool, lakeside climate of the Finger Lakes offers tremendous grape diversity, which no doubt contributes to its status as New York's largest wine-growing region. The sunny hillside slopes, moderating effects of the regional lakes, and relatively decent growing season combine to create world-class wines, particularly Riesling. Finger Lakes wines tend be a little leaner than their warm-weather cousins and often enjoy powerful aromas. Riesling and Gewürztraminer reach impressive aromatic heights within the region.

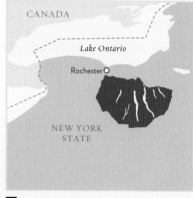

■ Finger Lakes region

Selecting and Pairing

Like many cool-climate grape-growing regions, the wines from the Finger Lakes pride themselves on well-honed acidity. Whether it's a cherry-laden Merlot, a well-weighted Cabernet Franc, or a lighter-styled Pinot Noir, you have many options for pairing with meat dishes. The fresh factor of Rieslings and clean, forward fruit of the local Chardonnay turn poultry and seafood dishes up a notch.

MEAT

Merlot, Cabernet Franc, and Pinot Noir befriend everything from pork tenderloin to prime rib, and sausage to stewed lamb. Grilled steaks, burgers, and pork chops find a friend in the versatile styles of the regional reds. Perhaps a bit unorthodox, Riesling also partners nicely with salty food finds, so you can give it a go with ham, charcuterie, and a variety of cold cuts.

CHEESE

The regional Cabernet Franc, Merlot, and Pinot Noir make mighty matches with Havarti, Gruyère, smoked cheddar, and Colby Jack cheeses.

POULTRY

Many off-dry Finger Lakes Rieslings work wonders with roasted poultry picks and really shine in the presence of Pacific Rim entrées filled with chicken, veggies, and a variety of Asian spices.

DESSERT

The Finger Lakes vineyards craft outstanding ice wines, often made from Riesling or Vidal grapes, which pair nicely with crème brûlée, shortbread cookies, vanilla-flavored cake, lemon meringue, peach cobbler, and a variety of fruit tarts.

SEAFOOD

Finger Lakes Chardonnay typically runs a bit leaner on the palate and works quite well with grilled or sautéed fish or shellfish (especially served in butter or cream sauce), and it brings out the best in crab cakes and clam chowder. A dry or semi-dry Riesling also pairs nicely with the range of flavors and cooking methods of the local seafood scene.

PRODUCERS TO TRY

Dr. Konstantin Frank, Fox Run, Hermann J. Wiemer, Heron Hill, Wagner Vineyards

UNITED STATES OF AMERICA

VIRGINIA

*Atlantic
Ocean*

WINES OF VIRGINIA

Close to 250 years ago, Thomas Jefferson, Virginia's original wine visionary, predicted that, "We could in the United States make as great a variety of wines as are made in Europe, not exactly of the same kinds, but doubtless as good." This was not a statement made lightly, given Jefferson's extensive experience with the finest of European wines. His own cellar records reveal a passion for the best of Bordeaux, Burgundy, Champagne, the Rhône Valley, Languedoc, Italy's Tuscany and Piedmont, German Riesling, Spain's Sherry, and a decent collection of Portugal's fortified finds.

Jefferson was truly America's first wine connoisseur. However, attempts to make wine in Virginia really began in the seventeenth century with the Jamestown settlement. Every able-bodied man was required to plant and cultivate a minimum of 10 European (*Vitis vinifera*) vines in an effort to make palatable wine. Unfortunately, their endeavors were crushed by phylloxera, a devastating, root-eating pest indigenous to the United States that attacked the root system of the European vines. In time, these European vines would be grafted onto native grape rootstock, mainly from the Norton grape, to overcome the persistent phylloxera and give the vines a fighting chance, even as molds, disease, and other climate-related hurdles continued to challenge vineyard efforts for years. In fact, it wasn't until the 1970s that Virginia's vineyards really began to take shape, and it's only in the past decade that Virginia's wines have been widely recognized for their quality and consistency.

Today, Virginia is home to more than 230 wineries and is America's fifth-largest wine-producing state, with 9 designated wine regions, hosting 7 registered American Viticultural Areas (AVAs): Monticello, Shenandoah, Eastern Shore, Northern Neck George Washington Birthplace, Rocky Knob, Middleburg, and North Fork of Roanoke.

Numerous grape varieties are grown in Virginia, but the most widely planted grapes continue to be Cabernet Franc, Cabernet Sauvignon, Merlot, Viognier, Chardonnay, and the native Norton variety, along with several French American hybrids.

More than two centuries later, Virginia has realized Jefferson's optimistic viticultural vision on many fronts.

Virginia's Monticello AVA

The Region and the Wine

Aptly named after Thomas Jefferson's remarkable home, the Monticello AVA hosts more than half of the state's wine grapes. The AVA extends to the base of the Blue Ridge Mountains, with some vineyards planted at more than 800 feet (244 meters); surrounds the charming city of Charlottesville; and covers 4 counties, 43 wineries (many of them boutique, family-owned), and more than 1,200 square miles (1,931 square kilometers). Key red grapes here are Cabernet Franc, Cabernet Sauvignon, and Merlot, with Viognier and Chardonnay covering the whites. The hardy, native Norton grape also keeps a moderate profile throughout the Monticello AVA.

Historically, Virginia has been a difficult place to grow wine grapes given the warm summer temperatures, high levels of humidity, harsh winters, and substantial potential for vineyard pests and diseases. However, significant efforts and education into site-specific vineyard management practices have paid off, allowing many of the region's grapes to thrive and ultimately reflect their own stunning sense of place.

Monticello region

Selecting and Pairing

The bright plum and cherry-driven profile of Cabernet Franc deserves a good deal of the regional pairing spotlight, with the fresh, lively flavors of Virginia's Viognier making a tasty splash with all sorts of seafood and poultry pairings.

MEAT

Cabernet Franc or regional Cabernet Sauvignon both work well with the full flavors of well-marbled cuts of beef, wild game, pork roasts, or the slightly sweet flavor of roasted lamb.

CHEESE

Cabernet Franc or Cabernet Sauvignon pair perfectly with the sturdy, sharp-natured cheese choices of cheddar, Gorgonzola, and Parmesan. For soft, creamy cheeses like Brie, Camembert, and some goat's and sheep's milk cheeses, the similar, softer textures of Chardonnay make a complementary pairing.

POULTRY

Local Chardonnay shows crisp acidity and the classic flavors of apple and pear with a touch of citrus, making it a tasty option for sautéed chicken; chicken Alfredo; or roasted, grilled, or baked game birds. With bright aromatics, exotic flavors, and a rich, round style, Viognier is a delight with a variety of poultry picks.

DESSERT

Virginia produces some exceptional dessert wines with intense aromas, a heavy fruit factor, and significant palate presence. These wines work with everything from crème brûlée, to poached pears, to fruit tarts.

SEAFOOD

Viognier is made for handling all sorts of seafood favorites, but it shines alongside local soft-shell crab dishes and flavorful crab cakes throughout seaside Virginia.

PRODUCERS TO TRY

Barboursville, Blenheim, Breaux, Chrysalis, Gabrielle Rause, Jefferson Vineyards, Keswick, Stinson, Sunset Hills, Veritas

PACIFIC
OCEAN

CANADA

Vancouver

UNITED STATES OF AMERICA

Montréal

ATLANTIC
OCEAN

WINES OF CANADA

Canada is known best for producing some of the world's top ice wines (referred to as one word, *icewine,* in Canada) made predominately from Riesling and Vidal grapes. Some adventurous producers also utilize the rich, red berry flavors of frozen Cabernet Franc. Table wines, sparkling wines, late harvest dessert wines, and ice wines are all produced within designated Canadian viticultural areas, although the sweeter side of the highly concentrated ice wines still claims the most accolades.

Canada's cool-weather climate grows an impressive variety of grapes, including many of the top international varieties of Riesling, Pinot Gris, Chardonnay, Gewürztraminer, Cabernet Franc, Merlot, Cabernet Sauvignon, and Pinot Noir, as well as some significant hybrids—Vidal being the most important.

Canadian table wines and sparkling wines can be difficult to find outside the country, but several major ice wine producers export their famous dessert wines in half bottles around the globe. Keep an eye out for Inniskillin, Jackson-Triggs, Pillitteri Estates, and Peller Estates. The high acidity and honeyed intensity of the region's ice wines, served well chilled, are palate-perfect for pairing with ripe, berry-themed desserts like angel food cake or shortbreads topped with raspberries, blueberries, or strawberries and cream, along with simple sugar cookies or cheesecake.

BRITISH COLUMBIA AND ONTARIO, CANADA

The Region and the Wine

Situated on opposite sides of Canada, British Columbia and Ontario are the country's most important wine-growing regions. Close proximity to large bodies of water are critical factors for their vineyard success, as the water provides a relatively warm buffer zone for the often-frigid Canadian climate. The vast majority of Ontario's vines skirt the Great Lakes, with the key appellations reflecting the famous waterways by name: Niagara Peninsula, Lake Erie North Shore, and Prince Edward County. British Columbia holds five designated growing regions but keeps most of its vines in the Okanagan Valley, where Okanagan Lake neighbors myriad microclimates, allowing vintners to experiment with a variety of grapes, landscapes, soils, and slopes, all in the context of dry, sunny days and cold, crisp evenings—perfect grape-growing conditions.

■ British Columbia
■ Ontario

Canada's growing regions cultivate more than 60 varieties of wine grapes. However, the grapes used for making the nation's renowned late harvest dessert and ice wines are of particular importance.

The late harvest wines are made from grapes that have been left on the vine to develop botrytis (a friendly fungus) that grows on the grape's skin and dehydrates the fruit to concentrate the sugar. Significantly less juice is available for pressing after botrytis has taken hold of a cluster of grapes, but what's pressed is an incredibly sweet nectar.

Ice wines are made in a different context. The grapes are frozen on the vines, hand-harvested, and then very gently pressed to separate the frozen water crystals from the prized, extremely concentrated sweet grape juice. The coveted juice is bottled in half-bottles, given the miniscule supply, and sold at premium prices.

The Vintners Quality Alliance (VQA) is the regulating body for Canadian wines. Similar to many other New World wine regions, a VQA wine must contain at least 85 percent of the grape variety featured on the label. It also must conform to specific regional growing standards that regulate allowed grape varieties, wine-making practices, and labeling laws.

PRODUCERS TO TRY

Henry of Pelam, Inniskillin, Peller Estates, Jackson-Triggs, Pillitteri Estates

WINES OF THE SOUTHERN HEMISPHERE

In the Southern Hemisphere, Australia, New Zealand, South America, and South Africa constitute the wine-making regions south of the equator.

Collectively considered New World wine regions, the compelling wines from these southern lands typically live up to their New World reputations, showcasing fresh, forward-fruit styles; bolder profiles; and higher levels of alcohol as many of their grapes get extended sun exposure and the potential to reach greater degrees of ripeness. The Southern Hemisphere grape harvest typically runs from early March to early June, depending on vintage variations—about the same time their Northern Hemisphere vineyard counterparts are experiencing spring's fervor of bud break and flowering.

Most wine-producing countries in the Southern Hemisphere owe their vineyard beginnings to European immigrants and missionaries who ventured forth to new frontiers. with vines in hand. Old World influences are still seen in the background practices of many southerly wine nations, yet in general, technology and innovation have laid much of the groundwork for the worldwide success and global demand of these exciting Southern Hemisphere wines.

Offering their own brand of adventure, education, and vinous personality, the wines from Australia, New Zealand, South America, and South Africa are particularly popular with the next generation of wine enthusiasts, who tend to closely follow ever-changing foodie trends and are often eager to share their latest wine discoveries via social media. The past two decades have seen a virtual wine revolution as the wine regions of the Southern Hemisphere—unencumbered by many of the Old World wine regulations that dictate vineyard yields, specific grapes for specific locations, and aging protocol—have experimented with a variety of vines, blends, and oak influences. Another factor that continues to impact the quality and eminence of many regional wines is the distinct, concentrating influence of increasingly older vines.

Wine fans who consider overall value, unique grape varieties, and easy access top priorities when buying wine will find that the Southern Hemisphere promises all that and much, much more.

INDONESIA

PAPUA NEW
GUINEA

Darwin ○

PACIFIC
OCEAN

AUSTRALIA

○ Brisbane

Perth
○

Sydney ○

○
CANBERRA

INDIAN OCEAN

TASMANIA

WINES OF AUSTRALIA

Australian wine is predominately associated with Shiraz (same grape, different name as Syrah), the big, friendly, berry-faced vino that's made plenty of friends worldwide with its amicable character, easy-on-the-wallet pricing, and super-tasty palate appeal. But Shiraz isn't all Australia's wine scene has to offer.

More than 100 different grape types are grown on this vast continent, with the bulk of the wines hailing from the state of South Australia and a handful of primary subregions: Adelaide Hills, the Barossa Valley, Clare Valley, Coonawarra, and McLaren Vale. Shiraz, Cabernet Sauvignon, Chardonnay, and Riesling tend to steal the limelight in South Australia.

New South Wales (home of the popular Hunter Valley) and neighboring Victoria constitute key appellations for Semillon, Chardonnay, and Shiraz.

The most notable wine region of Western Australia is Margaret River, 3 hours south of Perth, known for crafting brilliant Cabernet Sauvignon and world-class Chardonnay, with a gift for blending Sauvignon Blanc with Semillon to forge a wine with incredible class, complexity, and finish.

Technology, education, research, innovation, and automation are key factors in Australia's grape-growing, harvesting, and wine-making success. The generous use of screw caps to top Aussie wine is just one highly visible indicator of the country's forward-looking perspective on wine-making and wine-drinking.

Australia is a big wine exporter, and its wines trend toward the approachable, easygoing style that seems to personify many of the people who grow, harvest, blend, and bottle the region's inviting wines. Plenty of high-end Aussie wine options exist, but many down-to-earth producers seek to keep their prices in check and resist the temptation to charge exorbitant premiums, even on their prized wines—an approach many savvy consumers appreciate.

AUSTRALIAN WINE REGULATIONS

Australia's wine regulations are summed up in the Label Integrity Program (LIP), which is revisited and revised, if necessary, on an annual basis.

Essentially, the program requires Australian wine producers to adhere to a standard of blending and labeling rules. If a single grape is featured on the label, at least 85 percent of the wine in the bottle must be made from that grape and come from the printed vintage year. Also, 85 percent of the wine must be from the designated geographical area (referred to as *GI* for "geographical indication").

If the wine is a blend, things get tighter. The featured blended grapes must show percentages in descending order, and the blend must make up 85 percent of the labeled grape types.

In addition, 95 percent of the grapes must be from the labeled region, and all the grapes must have been harvested in the stated vintage year.

Southeast Australia is the wine region that graces bargain labels most frequently and encompasses a significant mass of vineyard land. Often, labels depict the producer, grapes, and the phrase *South Australia* or *Southeast Australia* to indicate that the grapes could be sourced from the continent's entire South or Southeast corner (South Australia, New South Wales, and Victoria), blending several regions together to build a consistent brand style that's not necessarily dependent on vintage variations.

Specific regional label designates like *Clare Valley, Hunter Valley,* and *Coonawarra* carry grapes that have been cultivated within significantly tighter boundary lines and tend to showcase more regional style components.

2011

BROKENWOOD

SEMILLON

Hunter Valley

750ml   WINE *of* AUSTRALIA   11,0% VOL

AUSSIE SALUTE 2011
WINE OF AUSTRALIA

d'Arenberg
ESTABLISHED 1912

The
Dead Arm

Shiraz

McLaren Vale
75cl   750mL

South Australia, New South Wales, Victoria, and Western Australia

The Region and the Wine

Australia is a big place, with lots of sun and warm, dry climate zones. Yet an abundance of biodiversity, geographical nuances, ocean influences, climate variations, soil substructures, and growing patterns weave throughout the nation's top wine-making regions. Tasmania, with its cooler climate, has been experimenting with the bright aromatics of Chardonnay and elegance of Pinot Noir with significant success, while the warm weather of the Barossa Valley brings juicy blackberry jam and blackcurrant fruit often accompanied by warm vanilla notes (and at times a touch of dark chocolate) to Shiraz.

Australia's regional differences create distinct palate variations. For example, it's interesting to taste how each appellation puts its own spin on Shiraz. The brawny, full-bodied, jam-filled Shiraz from old vines in the warmer Barossa Valley is quite different from the lower alcohol levels, overall elegance, and fine-structured style of Shiraz from the somewhat cooler climate of Clare Valley. Bone-dry Rieslings from Clare Valley contrast the sweeter side of the vine in Victoria's fortified wines. The lively Chardonnay, Pinot Noir, and even Shiraz-themed sparkling wines made in the traditional method of Champagne display an immense variety, from sweet to brut, and red to rosé.

Map legend:
- South Australia
- Western Australia
- Victoria
- New South Wales

Selecting and Pairing

With such dramatic wine diversity and multicultural culinary influences, pairing food with Australian wine is an adventure. Fresh produce, local seafood, and a variety of meat dishes top the menus of regional restaurants, all eager to match up with local wines.

MEAT

Shiraz is perfect for pizza, burgers, sausage (called *snags* Down Under), and barbecue. Local meat pies, sausage rolls, and roasted rack of lamb in red wine sauce are easy partners for the plush flavors and laid-back style of Shiraz. For up-scale roasts and choice beef cuts, opt for a fuller-bodied Shiraz, a Cabernet Sauvignon, or a blend of the two.

CHEESE

Creamy, soft-textured cheeses like Brie or Camembert make a delicious pairing with regional Chardonnay and also play well with a medium-bodied Cabernet Sauvignon. A well-aged Gouda or Edam welcomes a Shiraz.

POULTRY

Pheasant, duck, and barbecued chicken (dubbed *chook* regionally) take well to the rich fruit and soft tannins of Shiraz. Chardonnay is a top pick for chicken Dijon, creamy chicken and potato or pasta recipes, or turkey cold cuts, chicken salad sandwiches, and chicken curries (with unoaked versions).

DESSERT

Australia produces a surprising range of late harvest dessert and fortified wines (called *stickies* locally) that pair quite well with Lamington, a traditional Aussie treat made from sponge cake that's been dipped in a chocolate coating and topped with dried coconut flakes.

SEAFOOD

The regional John Dory pan-fried fish fillet topped with herbs or served alongside house-cut potatoes is a tasty pairing with the vibrancy of a dry Clare Valley Riesling or the citrus-driven acidity of Semillon. For crab cakes, fresh prawns, and lobster, scout for an Aussie Chardonnay.

PRODUCERS TO TRY

Brokenwood, Clarendon Hills, d'Arenberg, Greg Norman, Lindeman's, Penfolds, Penley Estate, Peter Lehmann, Rosemount, Shoofly, Tahlbik, Wolf Blass, Wynns, Yalumba

WINES OF NEW ZEALAND

New Zealand's reputation for adventure, sport, and adrenaline is captured in the dramatic backdrop of gentle hillsides sweeping into hulking mountain ranges; impressive glaciers giving rise to stunning fjords; and flat, fruit-filled plains rolling into rugged volcanic terrain, all of which is completely encircled by nearly 10,000 miles (16,093.5 kilometers) of shoreline. All this seems to somehow factor into the country's provocative, world-class Sauvignon Blanc wines, which willingly embrace their own standards of dazzling intensity.

New Zealand's climate is a mixture of plenty of sunshine followed by moderate rainfall and cooler temperatures, which allow New Zealand's grapes to ripen well and maintain extraordinary levels of food-friendly acidity.

New Zealand's Sauvignon Blanc is what planted the nation onto the world's esteemed wine map in the first place. Nowhere else has the Sauvignon Blanc grape been coaxed into showing such exquisite character, built on pure, often-exotic fruit; unmatched liveliness; and brilliant acidity. Tangy, pungent aromatics play toward the distinct smells of grapefruit and passion fruit, with the more subtle scents of peaches and pear often dancing in the background, joined at times by the warm, sweet scent of fresh-cut grass.

New Zealand is predominately white wine country. Chardonnay is the number-two grape, followed by an aromatic potpourri of Riesling, Gewürztraminer, and Pinot Gris under the white wine umbrella and Pinot Noir, Cabernet Sauvignon, and Syrah rounding out the top reds. In terms of numbers, New Zealand contributes less than 1 percent of the globe's wine, yet the wines it offers enjoy a significant following worldwide.

Stainless-steel fermentation tanks dominate the region's vinification process and shine the spotlight on the innate fruit character of the regional whites, with very little oak ever touching the grapes. New Zealand's wines are fresh, flavor-filled, and incredibly inviting. So it's no surprise Kiwis tend to be avid enthusiasts of screw caps to ensure the fresh fruit factors stay intact.

New Zealand wine regulations fall in line with much of the New World labeling themes. Essentially, the bottle must contain 85 percent of the juice from the stated grape variety, vintage year, and region listed on the label.

NEW ZEALAND'S NORTH ISLAND AND SOUTH ISLAND

The Region and the Wine

Divided into two major islands, aptly named the North Island and South Island, New Zealand features 10 main grape-growing regions. The warmer North Island hosts the primary regions of Hawke's Bay, Martinborough (technically a part of Wairarapa), and Gisborne. Martinborough is gaining significant ground with its outstanding Pinot Noir production. Less than 20 miles (32 kilometers) across Cook Strait, the cooler climate of the South Island carries the key players of Marlborough, home to Sauvignon Blanc, and Central Otago, which dominates with Pinot Noir.

New Zealand bottles an extroverted, fruit-forward, vibrant expression of the dynamic, highly versatile food-pairing Sauvignon Blanc grape. New Zealand's Marlborough region hosts 80 percent of the Sauvignon Blanc plantings, with sunny days and cool, crisp nights laying the groundwork for slow, even ripening that retains the grape's zingy acidity and results in wines of sustained intensity and precocious character. Sauvignon Blanc continues to be the region's brightest wine star, but Pinot Noir has recently been getting some attention, too. The cool Continental climate conditions, well-drained soil, and hillside plantings of Central Otago have wooed this demanding, thin-skinned grape into an elegant, accessible red wine, showing fine tannin; well-developed fruit; and herbal, earthy nuances.

■ North Island
■ South Island

Selecting and Pairing

Two of the world's best food-pairing wines just happen to be New Zealand's top grapes—Sauvignon Blanc and Pinot Noir. With a significant Pacific Rim culinary influence and loads of fresh seafood, the sophisticated flavors and textures of the local dishes are willing partners for the full range of New Zealand wines.

MEAT

Lamb is a popular menu item in New Zealand, perfect for Pinot Noir's silky textures, soft tannin, ripe raspberry fruit, and earthy character. For heavier meats like ribs, steak, and sausage, opt for New Zealand's Cabernet Sauvignon or Syrah.

CHEESE

Tangy goat's cheese and feta mellow zesty Sauvignon Blanc and develop the softer nuances of the cheese. The satinlike textures and ripe, red fruit of Pinot Noir engage the complexities of goat's and feta cheese. Camembert, Brie, and Gruyère also play well with Sauvignon Blanc and Pinot Noir.

POULTRY

Roasted chicken, turkey, duck, and pheasant pair well with the plush fruit and softer profile of Pinot Noir. Fruity Sauvignon Blanc and unoaked Chardonnay shine next to roasted, sautéed, or fried poultry, with Sauvignon Blanc easily cozying up to hard-to-pair veggies like artichoke, asparagus, and even bell peppers, herbs, greens, and assorted salads.

DESSERT

The traditional New Zealand dessert *pavlova* flourishes alongside a late harvest dessert wine, as do crème brûlée, cheesecakes, and ripe berry tarts. For something sweet and simple, pair Pinot Noir with a bar of silky smooth milk chocolate.

SEAFOOD

The bright, citrusy flavors and well-honed acidity of the regional Sauvignon Blanc are primed for pairing with New Zealand's local seafood (*kaimoana* in Maori), including oysters, scallops, pāua (abalone), *pipi* (a local shellfish), marinated mussels, lobster, cod, snapper, flounder, and more.

PRODUCERS TO TRY

Brancott, Craggy Range, Cloudy Bay, Dog Point, Kim Crawford, Nautilus, Nobilo, Villa Maria

PACIFIC
OCEAN

PERU

BRAZIL

BOLIVIA

ARGENTINA

Santiago

CHILE

ATLANTIC
OCEAN

WINES OF CHILE

Chile is a spectacular strip of land bordering the western edge of South America. Long and lean, stretching roughly 2,700 miles (4,345 kilometers) north to south and averaging a mere 100 miles (161 kilometers) east to west, Chile hosts enormous geographical diversity. The Atacama Desert binds the north, Patagonian ice fields enclose the south, the Andes mountain range hem in the east, and the Pacific Ocean defines the western border.

The middle of the country, surrounding the capital city of Santiago, is home to exceptional grape-growing conditions with a mix of soil structures, plenty of sunshine, and well-routed irrigation systems. Chile's unique physical isolation provides the region with a barrier to many modern-day vineyard pests, molds, and diseases. In fact, many believe that Chile (along with Argentina) escaped the worldwide phylloxera (a root-eating pest) epidemic that decimated many vineyards in the nineteenth century, thanks in part to its daunting physical barricades. With many of the typical vineyard pests held at bay, an increasing number of Chilean producers are targeting the organic wine market as part of their ambitious portfolios.

The Denomination of Origin (D.O.) is Chile's appellation system that divides the wine-growing areas into distinct "label-ready" regions. Chilean label requirements stick to the basics: if the wine bottle lists a grape type on the label, it must contain at least 85 percent of that grape. The same goes for the vintage year and growing region listed: 85 percent of the grapes must come from the vintage date and stated region.

Other than that, winemakers are fairly free to experiment with a variety of grapes, barrels, traditions, and technologies—the result of which has been an astonishing spike in Chilean wine quality and value over the last two decades, reaching both the budget buyer and collector on each end of the wine-buying spectrum.

Significant foreign wine-making interest, predominately from Italy, France, Spain, and the United States, has also supplied an infusion of capital, wine-making resources, and technology to many of the region's notable estates.

CHILE'S CENTRAL AND CASABLANCA VALLEYS

The Region and the Wine

Chile is home to more than 300 wineries and currently includes 14 designated wine regions, the most notable being the warm Central Valley (which includes the Maipo and Colchagua Valleys) and the cool, coastal Casablanca Valley. The Central Valley hosts the majority of Chile's vineyards and tends to focus on the distinct Chilean red wine grape Carmenère as well as Cabernet Sauvignon, Merlot, Syrah, Chardonnay, and Sauvignon Blanc. The foggy mornings and significant maritime influence of the Casablanca Valley are raising gorgeous cool-climate vines with an emphasis on crisp, fresh white wines (mainly Chardonnay and Sauvignon Blanc) and more recently, the demanding but delicious Pinot Noir.

■ Central and Casablanca Valleys

Chile offers an incredible range of microclimates, unique soil conditions, and assorted altitudes in its Mediterranean-like environment, and the variety of terroir creates a dynamic background for a mixture of grapes. The spicy, black-fruited, cocoa-themed, smooth-tannin character of Carmenère is Chile's signature red grape. Often blended, Chilean Carmenère's softer tannins find additional strength and structure alongside Cabernet Sauvignon. On the white wine front, the crisp, citrus-laced, often herbal character of Sauvignon Blanc makes for easy food-pairing versatility, while the round, rich structure of Chile's Chardonnay often exudes dense tropical notes and creamy textures.

Selecting and Pairing

Chile is a grower's paradise, and not just for grapes. Fresh fruit, fresh seafood, and regional meats find their way into local fare on a daily basis. Chilean wines successfully pair with an ever-increasing repertoire of dishes, and plenty of food-friendly wines are available at virtually all price points.

MEAT

A Carmenère–Cabernet Sauvignon blend is a top pick to pair with beef, inviting the spicy character, integrated tannin, and ripe fruit flavors to work their magic on everything from braised beef to filet mignon and grilled game to pork chops.

CHEESE

For breaded, herbed, or plain goat's cheese, there's no better choice than the herbal undertones of a lively Sauvignon Blanc. Blue, Brie, cheddar, and Camembert call for the well-defined structure, rich fruit flavors, and fuller-bodied profile of a Chilean Cabernet Sauvignon.

POULTRY

The rich, often-tropical character of Chilean Chardonnay provides a delicious backdrop for chicken smothered in cream sauce or simple chicken risotto. The refreshing acidity of a regional Sauvignon Blanc creates a perfect partner for chicken salad or recipes that call for fresh herbs and sautéed or roasted poultry.

DESSERT

Late harvest dessert wines from Chile aren't overly common, but several gutsy producers make some stunning late harvest treats from Sauvignon Blanc or Gewürztraminer grapes. If you find a late harvest Chilean wine, try it with ripe fruit tarts or pies, assorted cakes (including cheesecake), bread puddings, or shortbreads.

SEAFOOD

Raw oysters, well-seasoned grilled shrimp topped with lemon, a variety of fresh fish dishes, and a slew of spicy seafood stews welcome the zesty acidity, intense aromas, and notes of citrus, green apple, and Anjou pear that are often found in the palate mix of a well-grounded, Chilean Sauvignon Blanc.

PRODUCERS TO TRY

Arboleda, Casa Silva, Concha ya Toro, Cono Sur, De Martino, Emiliana, Errazuriz, Haras, Lapostolle, Montes, Primus, Santa Rita, Tarapacá, Veramonte, Viu Manent

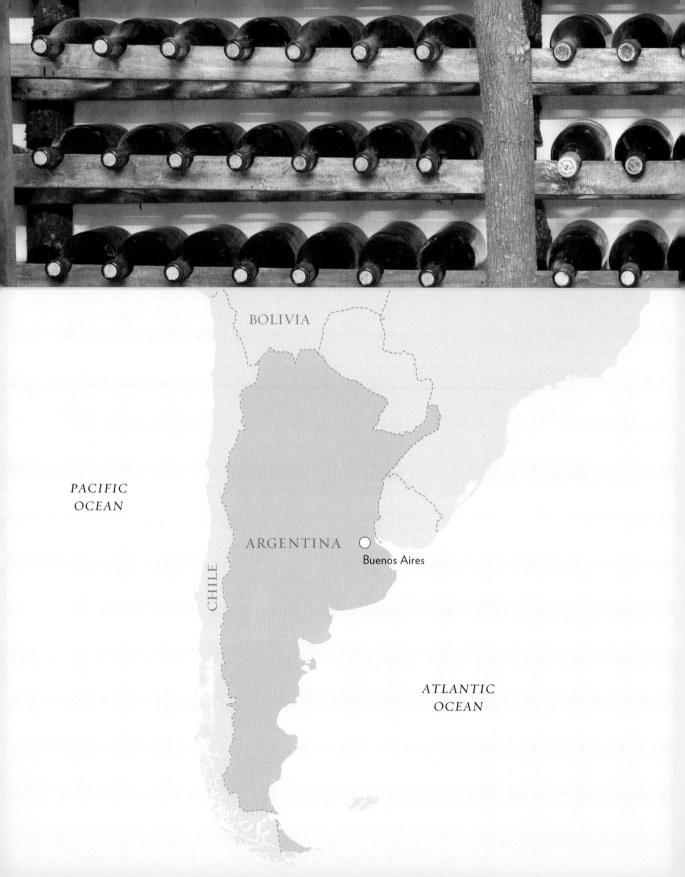

BOLIVIA

PACIFIC
OCEAN

ARGENTINA

CHILE

Buenos Aires

ATLANTIC
OCEAN

WINES OF ARGENTINA

Argentina is South America's largest wine-producing region and ranks fifth for overall global production. The relatively high altitudes, continental climate, and Andes Mountain water irrigation systems create a unique grape-growing environment, with few vineyard pests and diseases.

Red wines make up half of Argentina's wine scene, with the rich, red fruit; hefty tannins; and dense, velvety texture of Malbec leading the charge. The medium-acidity, zesty white fruit, and delicate floral profile of the regional Torrontés play the vibrant yin to Malbec's yang. Together these complementary wine counterparts act as some of Argentina's top international grape ambassadors.

Argentina has been growing grapes since the mid-1500s, when Spanish missionaries planted the first vines for Mass. This initial cultivation was followed by massive immigration from key European growing regions—mainly Italy, France, and Spain in the 1800s—many of which brought their regional, hometown vines with them.

Today, Argentina has experienced a virtual wine revolution. The last three decades have witnessed dramatic changes in quality and modernization as input, investment, and innovation from key wine-producing countries like France and the United States have sought to take the grapes, vines, and ultimately the wines of Argentina to world-class status. Although much of Argentina's export market has focused on solid-quality, value-priced wines, the high-end market category has shown considerable growth, proving that Argentina can meet market demands from the full spectrum of eager consumers.

MENDOZA, ARGENTINA

The Region and the Wine

Argentina is the fifth-largest wine-producing country in the world, with the majority of the country's exports landing in U.S. markets. Tucked into the base of the Andes Mountains, 70 percent of Argentina's vineyards are concentrated around the high-altitude hills and plateaus of Mendoza, with Maipú and San Rafael rounding out the significant subregions. Full-bodied Chardonnay and highly aromatic Torrontés are the key white wine players, with the bold flavors of Cabernet Sauvignon and the rich, velvety character of Malbec rounding out Argentina's dominant red wine grapes.

Mendoza region

Marked by an arid climate with significant altitude to balance the potential for extreme heat and promise plenty of sunshine, Argentina's vineyards thrive through warm, fruit-ripening days and cool, acid-balancing nights. The low humidity and natural barriers of the Andes to the west, the Atlantic to the east, and the Patagonian ice fields to the south provide physical protection against many vineyard diseases and pests. These ideal climate and growing conditions, along with good soil composition, make Argentina a prime location to grow world-class grapes. The last two decades have seen a surge in quality for Argentina's wines, thanks to significant international investments in both capital and wine-making expertise. Malbec is the nation's brightest wine star today, with international acclaim boosting sales worldwide.

Selecting and Pairing

Malbec is a mainstay for pairing with the country's burgeoning beef exportation market, providing willing tannin structure and pronounced dark fruit character. Seafood and poultry dishes delight in the floral aromatics and spicy nature of Torrontés as do exotic recipes from the Pacific Rim and regional empanadas.

MEAT

Beef is big business in Argentina, so it's no surprise Argentina's dominant red wine varietals partner well with red meat. Full-bodied wines with heavier tannin profiles and bold flavor are natural pairing partners for beef. The region's Malbec, Cabernet Sauvignon, and Syrah are ideal alongside roasted, stewed, and grilled beef.

CHEESE

With Brie, Camembert, Gruyère, Jarlsberg, and provolone, try a round, full-bodied Chardonnay. Stronger-flavored cheeses like cheddar, Danish blue, and Gouda work well with the big, bold profile of a Cabernet Sauvignon. The dense complexity of Malbec is ideal for partnering with a sheep's milk cheese like Manchego.

POULTRY

In general, chicken works well with the rich flavors and creamy textures of a medium- to full-bodied Chardonnay. The more cream and butter used in a dish, the better an oaked Chardonnay pairs. If the chicken is served in a lighter style, with fruit or herb accents, opt for the spicy aromatics and crisp acidity of Argentina's signature white wine grape, Torrontés.

DESSERT

Wines with higher levels of residual sugar tend to partner well with desserts. A late harvest or fortified Malbec works nicely with dark chocolate creations and berry-based desserts. For crème brûlée, almond-inspired desserts, and traditional fruit pies, go for a late harvest Torrontés.

SEAFOOD

The tangy acidity and zippy citrus flavors of lime and grapefruit that often mark Torrontés make it a perfect partner for a variety of shellfish dishes. Chardonnay provides both versatility and flavor appeal for seafood dishes that lean heavily on butter or cream-based sauces.

PRODUCERS TO TRY

Alamos, Antigal, Catena, Clos de los Siete, Hermanos, Luca, Michel Torino, Salentein, Susana Balbo, Trapiche, Trumpeter, Vina Cobos

AFRICA

NORTHERN CAPE

St Helena
Bay

WESTERN CAPE

O Prince Albert

EASTER
CAPE

Little Karoo

CAPE TOWN O

ATLANTIC OCEAN

WINES OF SOUTH AFRICA

South African wines present an interesting mix of New World and Old World styles. Although they're categorized and sold most often under the New World umbrella, a distinct Old World influence makes its subtle mark on various still, sparkling, fortified, and dessert wines. Combining the forward-fruit styles of many New World wine-producing regions with an earthbound, refined character that bring both Old World subtlety and easy elegance to many of the regional wines, South Africa often blends the best of both worlds in a single bottle. Reds, whites, rosés, sparkling wines, dessert wines, and fortified wines all populate the diverse South African wine portfolio.

Grapes have been grown here for several hundred years, starting with Dutch colonists, but South Africa's wine scene was largely built on bulk wine during the 1900s, produced in cooperatives with little or no character to speak of. In the background, private estates were producing quality wine, but due to international trade sanctions, the wine wasn't exported. Change came in the mid-1990s as politics shifted, apartheid ended, and Nelson Mandela became president. Long-standing international trade sanctions were lifted, and South African wines were shared with the world at large.

In the mid-1970s, South Africa established its Wines of Origin (WO) classification system. Considerably less complicated than the European Union's wine regulation laws, the WO designates how wine regions are defined from general geographic areas, to smaller delineated regions, to specific growing districts, as well as wine labeling requirements. South African wine law is nowhere near as strict as European regulation, but it does insist that regional locations adorn the labels, and if a grape or year is listed, at least 85 percent of the wine must be made with the specified grape and vintage year.

COASTAL SOUTH AFRICA

The Region and the Wine

The southern tip of Africa, buffeted between the Atlantic and Indian Oceans, enjoys more than 1,500 miles (2,414 kilometers) of coastline, a fairly temperate climate, and incredibly diverse grape-growing terroir. South Africa's most notable growing region, the Coastal Region, sits within the Western Cape, wraps around Cape Town, and extends north. The sandy and red clay soils of the Coastal Region contain several key districts: Stellenbosch (the largest of the three), Constantia (the oldest), and Paarl. The Coastal Region produces the vast majority of the nation's exported wine and hosts extraordinary diversity in both grape-growing varieties and sustained microclimates.

As the world's seventh-largest wine-producing country, South Africa cultivates a healthy mix of both red and white wine grapes. The smoky, spicy presence of Pinotage is perhaps the most well-known regional red, with Syrah and Cabernet Sauvignon rounding out the other top reds. The fresh-cut grass and distinct herbal character of the region's zesty Sauvignon Blanc has commanded considerable attention in recent vintages, although Chenin Blanc (dubbed *Steen* locally) has historically been the regional white wine star.

■ Coastal South Africa

Selecting and Pairing

South Africa has more than its fair share of wild game and seafood gracing local menus, and the international style of many South African wines make them an adventurous choice for pairing with a wide variety of flavors, foods, and cooking styles.

MEAT

Perfect for pairing with earthy wild game and the flavorful, sweet spice of barbecued beef, traditional burgers, or grilled brats and a daring delight with a meat-lover's pizza, the rustic, smoky flavors of a ripe, berry-flavored, medium-bodied Pinotage pairs well with a wide range of meats.

CHEESE

For a versatile cheese pairing, check out a South African red blend, leaning heavily on Syrah or Cabernet Sauvignon to create savory complements with cheddar, Brie, Camembert, Parmesan, or aged Gouda.

POULTRY

The steely, fresh green flavors of South African Sauvignon Blanc carry enough zesty, high-powered acidity to tackle virtually any poultry dish. Perfect partners include fried chicken, rice recipes with chicken and grilled vegetables, or roasted turkey with herb-filled stuffing.

DESSERT

Reputed to be one of Napoleon's favorites, South Africa's Constantia is a delicious, late harvest dessert wine, perfect for pairing with fresh fruit desserts, creamy custards, and caramel.

SEAFOOD

Quite at home with the abundance of local seafood in South Africa, the crisp, clean character of Chenin Blanc takes exceptionally well to oysters, mussels, lobster, crab, abalone, fried calamari, and grilled fish. Topping your favorite shellfish fare or fish dish with lots of lemon just boosts the experience.

PRODUCERS TO TRY

Bayten, Excelsior, Fairview, Glenelly, Indaba, Jam Jar, Kanonkop, Ken Forrester, Mulderbosch, Sutherland, Thelema

OTHER TYPES OF WINE

Most of the world's wines are red and white wine offerings, but there's a glorious realm of wine that goes far beyond the still, table wine arena. Champagne and sparkling wines celebrate everyday achievements or momentous milestones.

Fortified wines come in all kinds of shapes, sizes, and styles, made from fortifying a still table wine—red as well as white—with a neutral grape spirit—often brandy—and aged in various configurations, temperatures, and casks to create Sherry, Port, and Madeira. Today's fortified wine favorites please palates with their uncanny ability to pair with everything from tapas to crème brûlée.

Most of the dashing dessert wines, particularly late harvest and ice wines, are made possible by the somewhat unorthodox use of grapes harvested considerably later in the season, often with fuzzy fungus or full-on freezing frost onboard that concentrate flavors and intensify the sugars. Typically sold in smaller half bottles due to the extremely limited supply of sweet juice, it's estimated that the average late harvest or ice wine vine only produces enough sweet nectar to fill a single glass of wine, hence the price.

Whether it's sparkling wines and Champagne celebrations, full-bodied fortified wines built for winter warming, or delicious dessert wines that bring a sweet treat, the world of wine extends outside the familiar faces, names, and brands of everyday red, white, and rosé wines.

CHAMPAGNE AND SPARKLING WINE

Often reserved for special occasions and milestone celebrations, the festive bubbles of Champagne and sparkling wine make them extraordinary wines to star alongside food, with remarkable acidity, a vibrant and exceedingly versatile food-pairing nature, and relatively easy access worldwide.

Technically, Champagne is only Champagne when it's produced in the established region of Champagne, France, less than 100 miles (161 kilometers) northeast of Paris. All other bubbly falls distinctly into the "sparkling wine" category, often carrying a regional name like *Cava* for Spain's sparklers, various forms of *Spumante* or *Prosecco* for Italian bubbly, *Sekt* for Germany, and *Crémant* for everywhere else in France (as in *Crémant de Loire* for sparkling wines from the Loire).

Often made from a significant blend of vineyards, both Champagne and sparkling wine are adored for their effervescent bubbles, which are made by running the still wine through a secondary fermentation process to create and capture carbon dioxide. A Champagne or sparkling wine's final style depends on the country of origin, vintage variations, grapes used in the blend, the weight of the bubbly (light, medium, or full bodied), and varying degrees of residual sugar.

With so many styles, price points, straightforward pairing potentials, and jovial bubbles to top them off, it's no wonder today's Champagnes and sparkling wines are a hit wherever they're served.

CHAMPAGNE, FRANCE

Champagne takes the title as the most northerly wine-growing region in all of France, sitting just an hour and a half northeast of Paris by car or 45 minutes by train. Adored and admired around the globe for producing spectacular sparkling wine, the strictly defined region of Champagne is the only place in the world that grows, ferments, and bottles authentic Champagne. European Union regulations dictate that Champagne is only Champagne when it comes from the uniquely delineated AOC of Champagne, France, leaving little room (or patience) for international imposters.

With an absence of daily sunshine and consistent, cool-climate growing seasons, the grapes of Champagne are typically harvested starting in mid-September and have remarkably high levels of acidity and low levels of sugar, which could result in fairly insipid still wine. When forged into the world's most prestigious bubbly, however, this innate acidity results in an elegant sparkling wine with infinite complexity, exquisite taste, and an enduring full-flavor finish.

Many of the famous vineyards and Champagne houses—like Mumm, Moët and Chandon, Krug, Taittinger, Louis Roederer, Pommery, Pol Roger, Perrier-Jouët, Veuve Clicquot, and Charles Heidsieck—congregate around the history-rich towns of Rheims and Épernay. Three key growing regions exert significant influence: the Côte des Blancs (literally "the hillside of whites" for the predominance of Chardonnay grapes grown in the region's chalky soil) just south of Épernay; the Montagne de Reims (meaning "the mountain of Reims," known for cultivating exceptional Pinot Noir); and the Vallée de la Marne ("Valley of the Marne," named for the Marne River), which exhibits more sand and clay in the alluvial soil structure and tends to grow Pinot Meunier well.

The AOC of Champagne covers more than 80,000 acres (more than 32,000 hectares) of vineyard, with around 15,000 grape growers and hundreds of Champagne houses both big and small working together to produce well over 300 million bottles of Champagne annually, representing only around 10 percent of the world's sparkling wine market. (The United Kingdom is the largest export market.)

CHAMPAGNE GRAPES AND WINE LAWS

French law dictates that Champagne must be made from Chardonnay, Pinot Noir, and Pinot Meunier grapes. Pinot Noir and Pinot Meunier, both red wine grapes, account for well over half of the grapes cultivated in the region. Pinot Noir brings a bottle of Champagne more body, rich fruit aromas, heavier textures, solid structure, and additional power. Pinot Meunier brings fruity qualities, a round profile, and a supple character and contributes to the overall aromatics and earthy nuances. Chardonnay is an extremely important grape in the region, showcasing elegance and finesse, rich minerality, and lovely aromas.

The different ratios and proportions of the specific grapes used are up to the producers (called "houses" in France). However, Champagne is one of the most tightly regulated wine regions in the European Union, with specific standards set for vineyard classifications, vineyard yields, specific pruning heights and vine density measures, harvest limits, hand-harvesting requirements, wine-making, and aging process stipulations, among other particular details.

It's also interesting to note that most Champagne houses do not own their own vineyards. Instead, they source their grapes from growers who maintain long-term contracts with the individual houses. The relationship a Champagne house has with its vineyards, grapes, growers, and cooperatives are all encoded on the bottle labels. Two-letter bottle abbreviations offer specific insight.

Champagne Label Tips

Designate	What It Means
NM	*Négociant-manipulant;* means a Champagne house buys grapes to make its bubbly instead of growing them
CM	*Coopérative de manipulation;* indicates that a cooperative of growers has come together to make wines from their collective grapes
RM	*Récoltant-manipulant;* a grower who cultivates, harvests, presses, ferments, and ages his or her own wine to sell (also referred to as *grower Champagnes*)
RC	*Récoltant coopérateur;* a grower who cultivates and harvests his own grapes but relies on the cooperative to make the wine
MA	*Marque auxiliaire;* owned by a third party, like a supermarket, and not under the ownership of a grower, cooperative, or specific Champagne house

ABOUT THE BUBBLES

Essentially, a sparkling wine or Champagne is made from a still table wine that's put through a secondary fermentation. The various still wines that comprise the final blend of a sparkling wine or Champagne initially undergo the usual postharvest process. The grapes are picked, gently crushed, fermented to allow yeast to convert the sugar into alcohol and carbon dioxide, and often aged in barrel.

At this point, however, the process for creating Champagne (and sparkling wine) changes, as blends of various still wines (frequently from different vintages, unless of course it's a "vintage" Champagne) are blended together to build a stellar base wine. Factors affecting a Champagne blend include grape proportions; the amounts of Chardonnay, Pinot Noir, and Pinot Meunier used in the blend; grape origins (which vineyards from which villages); and grape vintage (which years work best in the blend). Some estimates point to an average of 45 different wines being used to form a blende d base wine in a typical bottle of Champagne.

Once the base wine is determined, it's run through a second fermentation process to create and capture the bubbles Champagnes and sparkling wines are known for. There are two ways this secondary fermentation takes place: the expensive, time-consuming method, or the quick and considerably cheaper alternative.

The Champagne Method

The costly and prolonged process of secondary fermentation is called the traditional method, the classic method, or the *méthode champenoise* ("Champagne method"). Not surprisingly, this is the one and only technique used for making genuine Champagne in Champagne, France.

The second fermentation takes place in the bottle. Starting with a well-blended base wine, a mixture of sugar and yeast (called a *liqueur de tirage*) is added to each bottle to kick off another round of fermentation—only this time, the bottle is capped and the carbon dioxide is trapped inside, creating Champagne's signature bubbles. This fermentation process takes a bit of time to get going and results in the formation of a layer of spent yeast cells called lees in the bottom of the bottle.

The longer the wine rests on the lees, often the more complex the wine's character can become. The yeasty aromas and creamy textures associated with many quality Champagnes are derived from the lingering influence of the lees, with top Champagnes often "resting on the lees," or *sur lie*, for several years, and sometimes more. Anywhere from a year to several years may pass before the dead yeast cells are removed from the bottle. During the extraction process, called *riddling*, the bottle is gently turned and tilted into a neck-down position so the lees sediment collects in the neck of the bottle. The bottle neck is then quickly frozen (termed *disgorgement* or *dégorgement*), the wine's cap is removed, and the lees-laden ice cube that's formed is popped out and discarded.

At this point, just prior to corking, adjustments to sweetness can be made through *dosage,* a technique that adds a tiny amount of sweetened wine to balance the naturally high acidity. Recently, an increasing number of Champagne houses are forgoing the *dosage*—usually when the wines are made from riper grapes—and labeling them *nondosé, zero dosage,* or *brut nature,* often to spotlight the intrinsic minerality completely unobstructed by additional sugar.

After the dosage decision is made, the bottles are corked and then readied for additional bottle aging—15 months minimum for nonvintage Champagne and 3 years minimum for vintage Champagne. Overall, in the traditional Champagne method, the wines enjoy amplified aromas, ongoing elegance, and finesse; the bubbles tend to be smaller, more precise, and more numerous; textures are often creamier; and the finish is more persistent.

The Charmat or Tank Method

The quicker and more cost-conscious method for running a base wine through a secondary fermentation is called the tank or Charmat method. Typically, the still wine is placed in a large, bulk tank and another round of sugar and yeast is added to initiate the second round of fermentation. The carbon dioxide is trapped, but this time it's in a temperature-controlled, pressurized fermentation tank, not a bottle, with a sealable lid and chilly temperatures.

The dead yeast, or lees, is filtered out of the wine under pressure to keep the bubbles in the tank, and if the sparkling wine needs any final tweaks in terms of sweetness, those are made in the tank, too, prior to bottling the bubbly. Instead of a year to several years, the Charmat method usually takes only weeks to a few months.

This Charmat process usually results in sparkling wines that showcase more fruit character on the palate compared to the Champagne method, where the primary influence is spent lees, which dials down the fruit and increases complexity.

"I drink champagne when I'm happy and when I'm sad. Sometimes I drink it when I'm alone. When I have company I consider it obligatory. I trifle with it if I'm not hungry and drink it when I am. Otherwise I never touch it—unless I'm thirsty."

—Madame Lilly Bollinger (Bollinger Champagne house)

CHAMPAGNE CATEGORIES AND STYLES

Grape types, blending proportions, and cluster quality, along with vintage or nonvintage designations and the varying degrees of dry to sweet, all merge to ultimately determine a Champagne's final aromatic, palate, and food-pairing expression. Whether the wine leans more heavily on the white wine grapes of Chardonnay, often resulting in a lighter body, or incorporates more of the region's red wine grapes, with the fresh fruit and zippy acidity of Pinot Noir or the earth-driven character of Pinot Meunier yielding a fuller body and greater intensity, the sheer variety of blends showcase the many categories and styles of world-class Champagne.

Types of Champagne

Type	Description
Blanc de blanc	Meaning "white [wine] from whites [grapes]," these fairly rare Champagnes are made from 100 percent Chardonnay grapes. Delicate and pale straw in color, often with hints of green, blanc de blancs promise elegance and finesse, with a light, dry palate profile. They also tend to be terrific with food.
Blanc de noirs	Meaning "white [wine] from black [grapes]," these Champagnes are made from 100 percent Pinot Noir or Pinot Meunier grapes. The color components of blanc de noirs range from deep golden to pale salmon, with a fuller body style than the blanc de blancs.
Rosé Champagne	Although not overly common and typically more expensive, rosé Champagnes are either made by simply adding still Pinot Noir to the wine prior to the secondary fermentation, which is the most common method, or by allowing the base wine to come into contact with the red grape skins until enough color has been extracted to give the wine a brilliant pink glow.

Vintage Versus Nonvintage Champagne

Champagne is built on blends. Blending grapes, villages, vineyards, and vintages into various sparkling wines that range from high-end luxury prestige *cuvées* and vintage bottles to everyday celebrations in a nonvintage format is precisely what distinguishes Champagne from virtually every other wine on the planet.

Nonvintage Champagne (typically abbreviated NV) is made by blending a minimum of two harvest years (and often more), with the majority of the base wine coming from the current harvest and about 30 percent from earlier vintages. In general, nonvintage Champagnes are dominated by the red wine grapes of Pinot Noir and Pinot Meunier and are rounded out by lower percentages of Chardonnay. These wines must be aged a minimum of 15 months in the bottle.

Nonvintage Champagne, typically the least-expensive option, is based on blending multiple grapes and multiple harvest years to magnify strengths, minimize weaknesses, and raise the bar on complexity and vintage-to-vintage consistency. The nonvintage wines comprise the bulk of the production and are often the best barometer for evaluating a Champagne house's prevailing style.

Vintage Champagne, as the name (and label) suggests, is made from grapes that come from a single harvest year, but the same full-throttle blending of grapes (although the earthy profile of Pinot Meunier is often abandoned for vintage Champagne), villages, and vineyards remain an integral part of the overall process. With vintage Champagne, however, the quality of the vineyards and grapes are turned up a notch, and the wine's profile typically exudes a fuller body, deeper complexity, and longer finish than its nonvintage Champagne cousins.

Due to Champagne's harsh weather conditions, vintage years only come around every 2, 3, or sometimes 4 years, when the weather cooperates and grapes are in top form. Individual Champagne houses determine whether or not the year is good enough to make vintage Champagne. If a house decides to go for it, it must age its wines a minimum of 3 years prior to release, but most go for more, with many houses doubling the minimum aging requirements and cellaring their wines for 6 years before release. Due to supply and demand, the limited production, and hefty aging stipulations, vintage Champagnes come with an understandably higher price tag.

Prestige cuvée, also known as premium vintage or *tête de cuvée* (literally, "head of vintage"), are high-end Champagnes that focus on being the best of the best. Utilizing the best grapes from the very best vineyards and often only sourcing from a single vintage, famous prestige cuvées include Dom Pérignon, La Grande Dame, and Cristal. This category represents the priciest Champagne bottles on the market, promising more concentration, greater intensity, and lofty label prestige.

Each Champagne house blends its bottles to produce a consistent house style. This house style is determined by the wine's body (light, medium, or full bodied), which is significantly influenced by the grape types chosen for the blend; how sweet or dry the wine presents itself; and whether the house is shooting for fruit or finesse, elegance, endurance, and complexity or lively acidity and delicate textures with fairly streamlined palate profiles. Producers intentionally blend for the same dependable style every year, and by combining a variety of grapes, vineyard locations, and vintage years, the goal is quite achievable.

General Champagne House Styles

Champagne houses pride themselves on being able to deliver a consistent house style year after year to consumers who count on a particular palate profile. Often the initial indicator of house style is found in the weight of the wine. Is it light bodied, delicate, and showing considerable elegance? Or does the wine offer a medium-bodied style with a bit more palate weight? Finally, which Champagne can you count on to offer a full-bodied, powerful style?

Different Champagne houses have established different house styles that have built both their reputation and position. These well-known Champagne houses also represent solid, well-distributed producers to try:

Light-bodied house styles:

> G.H. Mumm
>
> Laurent-Perrier
>
> Perrier-Jouët
>
> Pommery
>
> Taittinger

Medium-bodied house styles:

> Billecart-Salmon
>
> Charles Heidsieck
>
> Moët and Chandon
>
> Nicolas Feuillatte
>
> Piper-Heidsieck
>
> Pol Roger

Full-bodied house styles:

> Bollinger
>
> Krug
>
> Louis Roederer
>
> Veuve Clicquot

THE CHAMPAGNE SWEETNESS SPECTRUM

Until the mid-1800s, Champagne was crafted in a sweeter style than it is today, as certain Champagne houses started cutting back on the additional sugar, creating the first brut (dry) styles. Today, the majority of Champagne falls decidedly into the brut category.

The spectrum of sweet to dry is technically determined by Champagne's residual sugar levels—a result of the combination of wine and sugar added to the wine after the secondary fermentation known as *dosage*—each category along the spectrum offers a full range of Champagnes.

There are six important categories of Champagne to know; many of the world's sparkling wines follow the same terminology to communicate general levels of perceived sweetness:

Dosage is a sweetened still wine used to modify a sparkling wine's or Champagne's sugar structure, determining how sweet or how dry the final bottle of bubbly will be. The *dosage* is administered after the spent yeast sediment is removed from the bottle neck and just before corking.

Extra brut is extremely dry with sugar levels that fall into the 0 to .6 percent sugar per liter range (0 to 6 grams sugar per liter).

Brut is dry with sugar levels that fall below 1.5 percent sugar per liter (less than 15 grams sugar per liter).

Extra dry implies that these Champagnes would be drier than brut, but they're usually not. The sugar levels fall into the 1.2 to 2 percent sugar per liter range (12 to 20 grams sugar per liter). Champagne in this category is often served as an aperitif.

Sec, French for "dry" (although these Champagnes often present in a semi-sweet nature), sec Champagnes carry sugar levels that fall into the 1.7 to 3.5 percent sugar per liter range (17 to 35 grams sugar per liter).

Demi-sec, or literally "half-dry," Champagnes maintain sugar levels that fall into the 3.3 to 5 percent sugar per liter range (33 to 50 grams sugar per liter). This category is palate perfect for pairing with a variety of desserts.

Doux, or "sweet," Champagnes are very sweet and very rare. They hold a sugar level of 5 percent or more per liter (more than 50 grams sugar per liter).

A quick peek at the various categories of Champagne ranging from bone-dry to super sweet shows there can be considerable overlap within the *dosage* divisions. For example, one Champagne house might label its bubbly with 1.5 percent sugar (15 grams) per liter as brut, while another house with the exact same sugar level might label their wine conservatively as extra dry. Regional regulations define *extra dry* as a wine that has 1.2 to 2 percent sugar or 12 to 20 grams per liter and *brut* as a wine that weighs in at 1.5 percent or 15 grams (or less) sugar per liter.

There's a bit of wiggle room in terms of labeling and communication of the wine's style and degree of sweetness. That's precisely why many Champagne houses have consistent fans and followers who have become quite familiar with their particular house style and have come to rely on and expect similar styles from year to year.

CHAMPAGNE BOTTLE SIZES

The most common sizes of Champagne and sparkling wine bottles are the standard 750-milliliter bottle and the magnum, which contains the equivalent of two regular bottles and weighs in at 1.5 liters. You'll also find smaller split bottles (187.5 milliliters, or a little over a glass) and half-bottles (375 milliliters, or 2 glasses).

However, especially in Champagne, special hand-blown, oversize glass bottles are crafted to carry a wide range of bubbly. Although impressive in quantity, they're not overly common and are generally reserved for exceptional occasions. The lofty names of these bottles are suitably named after mostly biblical kings:

Jeroboam holds the equivalent of 4 bottles, 3 liters, or close to 20 glasses. Jeroboam was the first king of the northern kingdom of Israel after Israel divided into the two kingdoms of Israel and Judah. He reigned from 931 to 910 B.C.

Rehoboam holds the equivalent of 6 bottles, 4.5 liters, or close to 30 glasses. Rehoboam was the son of Solomon and king of Judah from 931 to 913 B.C.

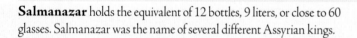

Methuselah holds the equivalent of 8 bottles, 6 liters, or close to 40 glasses. Methuselah wasn't a king, but he was a biblical patriarch, grandfather of Noah, and the oldest living man in the Bible at 969 years.

Salmanazar holds the equivalent of 12 bottles, 9 liters, or close to 60 glasses. Salmanazar was the name of several different Assyrian kings.

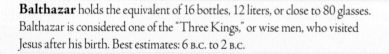

Balthazar holds the equivalent of 16 bottles, 12 liters, or close to 80 glasses. Balthazar is considered one of the "Three Kings," or wise men, who visited Jesus after his birth. Best estimates: 6 B.C. to 2 B.C.

Nebuchadnezzer holds the equivalent of 20 bottles, 15 liters, or close to 100 glasses. King of Babylon from 605 to 562 B.C., Nebuchadnezzer conquered Judah, destroyed Solomon's Temple in Jerusalem, and took captives to Babylon.

REGIONAL SPARKLING WINES

Sparkling wines made around the world tend to follow the classic model of Champagne as a starting point and then incorporate their own regional flare via grapes, locations, and style. Most major wine-growing regions make some form of sparkling wine, although quantities vary quite a bit.

For international sparkling wines, Spanish Cava and the Italian bubbles of Prosecco and semisparkling Moscato d'Asti typically top the charts when it comes to overall value and food-friendly features. The United States produces a number of dazzling sparkling wines from California, Washington, and New York. Smaller wine-producing nations like Germany and Austria produce locally beloved bubbly called *sekt,* with limited exportation and maximum local demand. France makes plenty of sparkling wine in regions outside of Champagne (typically with grapes beyond Champagne's big three), but they come under the name Crémant and tag the regional name on the end. For example, Crémant d'Alsace is a sparkling wine from the northeast region of Alsace, and Crémant de Limoux is a sparkler specifically from the southern French region of Limoux. Crémants are made in the traditional method of Champagne with heavy reliance on local grapes.

The Southern Hemisphere also produces its fair share of sparkling wine stars. Australia makes first-rate sparkling wine, some even from Shiraz grapes, with both the traditional and Charmat method of secondary fermentation. New Zealand creates delicious seafood-friendly sparkling wines, many in the traditional method, most often with classic Pinot Noir and Chardonnay grapes. South Africa and South America both export unique sparkling wines that capture the region's character, dominant grapes, and inviting styles.

International sparkling wines may shoot for similar grapes and styles as those that make the famous bubbly of Champagne, or they may showcase their own native grapes and opt for using the Charmat method for the second fermentation instead of the Champagne method (*méthode champenoise*). Yet many sparkling wine producers still use much of the same terminology for categorizing their sparkling wines based on Champagne. Using terms like *brut* or *extra brut;* adjusting sugar levels after the second fermentation; and crafting house styles from a series of blends based on grapes, vineyards, and vintages all remain commonplace for the global sparkling wine scene.

Spain's Spectacular Sparkler: Cava

Produced in the northwest corner of Spain, in the famed Penedès region just outside of Barcelona, Spanish Cava makes its mark as one of the wine world's most affordable sparkling wines.

Crafted using the traditional method of Champagne wherein the second fermentation takes place in the bottle, Cava is dependent upon the dynamic blend of several native white grape varieties—Macabeo, Parellada, and Xarel-lo—and the international Chardonnay.

Macabeo lends a fresh, fruity quality and delicate elegance to the sparkling wine, while Parellada contributes more concentrated aromas that often center around green apple and some citrus. Xarel-lo brings body and acidity and an earthy undertone, and Chardonnay offers body, aroma, and finesse. Cava delivers a crisp, lively sparkling wine made with dry to off-dry levels of sweetness typically in a light to medium body. Perfect for pairing with everything from goat's cheese and triple-cream Brie; to the salty nature of olives and nuts; to fresh shellfish, tuna, and salmon; other rich fare, whether fried or butter-laden; and even a bite of classic quiche lorraine—all delight in a Cava match.

PRODUCERS TO TRY

Castillo Perelada, Codorníu, Freixenet, Jaume Serra Cristalino, Juve y Camps, Loxarel, Raventós i Blanc, Segura Viudas

Italy's Popular Spumantes: Prosecco and Moscato d'Asti

Italians call their sparkling wines *spumante.* Spumantes may be made in both semisparkling or traditional sparkling varieties and span the spectrum from quite dry to considerably sweet.

Prosecco and Moscato d'Asti are the most well-known Italian bubblies, and both boast fresh, fruit character with delicate aromas, lighter body weight, and lively acidity. The stainless-steel Charmat method is the bubble-making process of choice for Italian sparklers, as it promises to promote the bright fruit factor, go fairly easy on the bubbles, and make an ultra-approachable summer sipper.

From Veneto's Friuli region, in the northwest corner of Italy, comes Prosecco, one of the most popular, wallet-friendly sparkling wines around made with grapes by the same name. Ranging from semisparkling to sparkling and generally falling into the dry or brut category, Prosecco may come across as slightly sweet due to the ambitious forward, fruit-filled character that tends to highlight apple, pear, citrus, and tropical fruit, all topped off with a dash of honey and cream. Prosecco wines offer a tremendous pairing with shellfish and have a distinct regional preference for clams and mussels. Fried fish and chips, spicy Pacific Rim cuisine, and almond desserts all can count on Prosecco's bright fruit flavors, light alcohol, and zesty acidity to make the very most of the match.

Moscato d'Asti hails from Italy's northeast Piedmont region and offers a delicate, slightly sparkling wine (dubbed *frizzante*) that's fairly inexpensive and extremely well received. Made in the Italian town of Asti from Moscato grapes, the lower alcohol levels, light-bodied style, splash of sweet, and engaging floral aromas bring an unmatched freshness to the table. Moscato d'Asti is well positioned to partner with all sorts of desserts, especially those that spotlight ripe fruit, cream, and cake.

PRODUCERS TO TRY

Prosecco:

La Marca, Lamberti, Nino Franco, Mionetto, Santa Margherita, Villa Sandi, Zardetto, Zonin

Moscato d'Asti:

Castello del Poggio, Ceretto, Chiarlo, Fratelli, Saracco

America's Sparkling Wines

America's sparkling wine market is strong and continues to gain momentum. J Vineyards, Iron Horse Vineyards, Handley Cellars, and Schramsberg in California, along with Domaine Ste. Michelle and Pacific Rim in Washington, and houses in New York, Oregon, Virginia, and New Mexico all produce delicious bubbly.

Many of the top French Champagne houses also have stakes in California's sparkling wine arena, including Domaine Chandon, a piece of the Moët and Chandon powerhouse; Domaine Carneros, a high-quality extension of Taittinger; Roederer Estate, part of the Louis Roederer portfolio; Mumm Napa, an offshoot of G.H. Mumm; or Pacific Echo, owned by Veuve Clicquot. All these estates represent important properties in California's sparkling wines, and all carry a decidedly French influence.

In the United States, especially in California, sparkling wine tends to be front-loaded with riper fruit than its Old World counterparts. The sunny days, often longer growing season, and vastly different terroir contrast the consistently cool-climate of Champagne, France, on almost every front. Grapes often find full, sun-driven maturity in the hills and flats of Napa and Sonoma, highlighting exceptional apple, citrus, and tropical nuances often wrapped in the restraint and elegance of Old World Champagne styles, thanks in part to the traditional bottle fermentation process. (If it's an inexpensive bottle of American bubbly, typically under $10/€7.50, the traditional method was not part of the process and the wine will exhibit even more fruit and less complexity.)

Food pairing for many of America's sparkling wines mirror the same themes as Champagne. Seafood, fried food, and egg-inspired dishes often top the pairing charts for bubbly with Asian dishes, cheese (especially Brie, Camembert, and Gruyère), various hors d'oeuvres, and smoked salmon or lox.

PRODUCERS TO TRY

Chateau Frank, Domaine Carneros, Domaine Chandon, Domaine Ste. Michelle, Gloria Ferrer, Gruet, Handley Cellars, Iron Horse Vineyards, J Vineyards, Mumm Napa, Roederer Estate, Schramsberg

STORING CHAMPAGNE AND SPARKLING WINE

Like their still wine counterparts, bottles of Champagne and sparkling wines do best when they're stored on their sides in a cool, dark environment. Keep in mind, however, that sparkling wines are even more sensitive to fluctuations in temperature and excessive light than typical still wines. Ideal storage temperatures run around 55°F (13°C), which is close to the natural temperatures of the underground cellars and caves where many of the region's producers age their bottles.

Both Champagne and sparkling wine should be served well chilled to highlight the aromas, flavors, and bubbles, but also to tone down some of the cork pressure prior to opening. The best way to chill a bottle of bubbly is to fill a bucket half full with ice, place the bottle in the ice, and fill the bucket the rest of the way (to the bottle neck) with cold water. This ice bath chills the sparkling wine well in 20 to 30 minutes.

OPENING CHAMPAGNE AND SPARKLING WINE

The average bottle of Champagne or sparkling wine carries close to 90 pounds per square inch of pressure (6 atmospheres of pressure)—double and even triple what many car tires hold and enough power to do some serious damage if the bottle is incorrectly uncorked. Some sparkling wine styles, including many French Crémants and Italian Proseccos, make bubbly with a bit less pressure.

The thicker bottles and fatter, caged corks used in Champagnes and sparkling wines are uniquely designed to deal with the intense pressure demands, but they also require some patience and practice when it comes to opening. Shoot for a solid 20 to 30 minutes in an ice water bath to ensure the wine reaches the ideal serving temperature of 45°F (8°C) and decreases the pressure on the cork.

It's essential to have the Champagne or sparkling wine bottle properly chilled prior to opening to keep the cork from launching and the bubbles from spilling.

Here's how to open a bottle of Champagne or sparkling wine:

1. After the bottle has been chilled, remove the foil capsule at the top of the bottle covering the cork.

2. Keeping your thumb on the cork, gently twist the wire ring attached to the wire cage that holds the top of the cork in place, and lift it off the top of the cork.

3. Angle the bottle at about 45 degrees, pointing away from people, pets, and breakable items like lights.

4. Carefully begin to ease the cork up and out of the bottle with your thumb while slightly twisting the base of the bottle in the opposite direction with your other hand.

5. When you feel the cork release from the neck of the bottle, it should be accompanied by a slight sigh, not a loud, wine-spilling pop. (Some people prefer to place a clean towel or cloth over the top of the bottle before easing out the cork to catch the cork in the cloth if it releases with a pop rather than a sigh.)

6. Slowly pour the Champagne or sparkling wines into flutes or tulip-shape glasses, but only about halfway full to allow for the fizz factor and keep from overflowing the glass.

The size and shape of the traditional Champagne flute and tulip-shape glassware concentrate both the aromas and bubbles, streamlining the route the latter take as they hastily make their way to the top of the glass.

FLAVOR AND FOOD PAIRING

Known for its bright acidity, thanks to the cool northerly climate, limited sunshine, chalky soils, and perpetual struggle to mature grapes, Champagne's distinct aromas, flavors, and textures reflect the many faces and facets of the region. The hallmark aromas often steer toward the alluring scents of fresh-baked bread with a toasty, yeast-filled profile, especially when made in the traditional method of Champagne, with the full range of apple aromas from classic, off-the-tree apple slices to the sweeter, more concentrated aromatics of applesauce. Citrus, poached pears, nuts, caramel, brioche, and fresh cream round out the familiar scents, tastes, and textures of many of today's popular Champagnes.

Given Champagne's and sparkling wine's affinity for food; its acid-driven nature; its lively notes of apple, citrus, bread dough, spice, and cream; and its often underlying minerality, the rich textures, festive bubbles, and ability to play well with a range of flavors, preparations, and individual preferences make it a top pick for everything from appetizers to dinner and even desserts, especially with the sweeter styles of demi-sec and doux. In fact, when several people at a table are ordering different menu options but would like to share a bottle of wine, sparkling wines are an easy and appealing pairing solution, thanks to their tremendous ability to please a variety of palates under diverse serving conditions.

Typically, the drier the Champagne (extra brut, brut, and extra dry), the better it handles as an aperitif or full-fledged meal partner. The sweeter-styled Champagnes (sec, demi-sec, and doux) shine alongside the likes of cheesecake, wedding cake, and fruit-themed desserts and pastries.

In general, a medium-bodied, dry Champagne or sparkling wine presents an astonishing résumé of favorite food pairings. From classic caviar, to calamari and butter-bathed lobster, to smoked salmon and oysters, to savory sushi and somewhat spicy Asian fare all the way to rich butter sauces, cream-themed recipes, egg-driven dishes, and fried feasts, the fat-cutting, flavor-enhancing acidity and palate-cleansing bubbles in Champagne and sparkling wine present an unbelievable partnership.

English writer Aldous Huxley had it right when he quipped, "Champagne has the taste of an apple peeled with a steel knife."

Although Champagne and sparkling wines are often presented as best for "formal fare," their extreme versatility is apparent even when matched with more casual cuisine. When paired with buttered popcorn, potato chips, French fries, and fried chicken, Champagne and sparkling wine shine with buttery aromas and crisp acidity and meld into a refreshing albeit unconventional, palate-pleasing union.

Pairing Champagne and Sparkling Wine Styles with Food

Style	Favorite Foods
Blanc de blanc	The 100 percent Chardonnay themes of blanc de blancs often bring a bright pear or crisp green apple aroma with a refreshing palate feel supported by a dose of creamy textures. Brie or fontina cheese, fresh cream-inspired dishes, savory canapés, caviar, seared scallops, oysters, and white truffles all play quite well with the profile of traditional blanc de blancs.
Blanc de noirs	Made only from Pinot Noir grapes, the sometimes earthy flavors, clean structure, and raspberry-inspired fruit of blanc de noirs partner well with everything from pork tenderloin with fresh rosemary and thyme to crusted halibut, smoked salmon, or roasted beef topped with mushroom sauce.
Rosé Champagne or sparkling wine	Perfect for serving with smoked salmon or tuna; roasted lamb or duck; fresh lobster; numerous pork dishes; and cherry, strawberry, and raspberry desserts, a sparkling rosé offers tremendous food-friendly partnerships.
Sec	This semisweet Champagne or sparkling wine finds its match in a variety of cheeses, fresh salads, and fruit-filled dishes.
Demi-sec	Delicious with dessert, the sweeter style of demi-sec brings bubbles to any sweet treat, particularly those featuring fresh fruit, pastries, or even some chocolate confections.
Doux	The sweetest sparkler of all, and often tough to find, doux Champagnes and sparkling wines successfully woo a wide range of desserts. From custards and caramels to fresh fruit and spicy pumpkin pies, the sweeter sparklers balance the acidity, concentration, and overall sweet factor with exceptional ease.

FORTIFIED WINES

Fortified wines are still wines that have been "fortified" by the addition of a neutral grape spirit that yields a wine with higher alcohol levels, typically in the 15 to 22 percent range. Neutral grape spirits are distilled from grape juice that's been fermented into wine. Brandy is one of the most well known. In fact, *brandy* is the abbreviated version of *brandywine,* which stems from the Dutch word *brandewijn* meaning "burnt wine."

Although many fortified wines come with a significant dose of residual sugar, thanks to early fortification that brings fermentation to a standstill and keeps the innate sugar from being converted to alcohol, a number of wines are fortified at the end of fermentation and show a completely dry style.

Most fortified wines find plenty of time for aging in wood, with several notable exceptions spending more time aging in bottle. The extra aging promises greater depth and complexity and opens new avenues for both secondary and tertiary aromatic character development.

The most famous fortified wine has to be Port, from northern Portugal, but plenty of fortified variations exist throughout the world's wine regions. Madeira is another fortified gem from Portugal, practically indestructible due to its heat-induced oxidation. Sherry, made predominately from Palomino grapes in the sun-kissed "Sherry triangle" of southern Spain, claims the limelight for Spain's fortified front. Italy's Marsala, made from native grapes, is similar to Sherry in that it's fortified after fermentation, comes in a variety of styles, and is blended and aged using a perpetuum system similar to Sherry's solera system. In France, Muscat de Beaumes de Venise represents a sweet, fortified, Muscat-based wine from the Rhône Valley (the village is Beaumes de Venise), and Banyuls from the south is one of the most popular French fortified *vins doux* (sweet wines) produced from regional Grenache grapes and pairs extraordinarily well with chocolate.

Fortified wines tend to be excellent food partners, offering a range of diversity from aperitif to dessert, depending on the specific style, age, and attributes of the wine. Many nuts, tapas, charcuterie, and appetizers work well with the drier fortified wines, while the sweet character of the dessert styles beg for fruit themes, custards, chocolate, and pastries.

SHERRY

Sherry, Spain's well-known fortified wine, is proudly made in the sunny, southwest Andalusia region. Bright, white sand and chalky, limestone-laden soils (locally dubbed *albariza*) form the foundation for cultivating the legacy of vines that connect the distinct landscape of the trio of primary Sherry-producing towns in Spain's proverbial "Sherry triangle"—Jerez de la Frontera, El Puerto de Santa María (also famous for being the port of departure for Columbus' second trip to the New World), and Sanlúcar de Barrameda. Coastlines, hillsides, and plenty of sunshine make their mark on the truly unique terroir sitting just west of the Strait of Gibraltar.

Only three white wine grapes are incorporated into the production of Sherry: the Palomino grape (by far the biggest contributor), Pedro Ximénez (a big player in dessert Sherry), and Moscatel (a.k.a. Muscat) to a lesser degree.

Sherry Styles

Ultimately, Sherry is a blend of multiple vintages, with every Sherry house maintaining and marketing its own signature style. Sherry is an extraordinarily diverse fortified wine, from the nimble, light-bodied, distinctly dry versions like fino and manzanilla; to the full-bodied, robust flavors of the oxidized styles of oloroso; and the darling of the dessert world, Pedro Ximénez.

The process of making Sherry begins the same as any still wine: grapes are grown, harvested, crushed, and fermented to create a base wine and then spend some time aging in a barrel prior to fortification.

The base wine is evaluated for aromas, body, and initial impressions and then fortified accordingly with a neutral grape spirit. The lighter, paler-colored base wines showing delicate aromas and flavors are fortified with lower doses of a neutral grape spirit, shooting for about 15 percent alcohol. Base wines with heavier bodies and powerful profiles are fortified to around 17 percent alcohol.

The Function of Flor

A lower dose of neutral grape spirit added to the base wine takes the Sherry-to-be down the lighter-styled fino and manzanilla road. The lower alcohol load permits an unusual layer of wild yeast, called *flor,* to essentially blanket the developing fino or manzanilla in the barrel and prevent significant oxidation.

With higher levels of a neutral grape spirit, the wild yeast is prevented from forming the layer of flor. Without the flor barrier, the fortified base wine undergoes some oxidation. The oloroso styles of Sherry (labeled most often as *oloroso, cream,* or *Pedro Ximénez*) rely on a bit of oxidation to produce a rich, full-bodied, character and give the wines an enduring edge. Once opened, they'll hold their own quite well for several weeks (when refrigerated), thanks to the early impact of intentional oxidation and higher levels of alcohol.

The Sherry Solera System

Sherry is a unique wine on many fronts, and its aging system is one more distinction on the route to bottling. The solera system is a progressive method of aging and blending Sherry that enables producers to maintain a consistent house style. Young wines are fractionally blended with older wines in a process that brings a variety of vintages into a single bottle.

Multiple rows of barrels filled with Sherry at various levels of maturity are categorized by age. The barrels holding the oldest Sherry are partially drawn off—not more than a third of the barrel—to bottle and then refilled by the next oldest barrels, which in turn are refilled by the next oldest barrels, until the recently fortified Sherries are reached and blended with slightly older barrels on their trek to the bottom level of older barrels.

The initial vintage plays a part in the grapes of the base wine, but the final Sherry bottle holds a panoramic blend of many vintages and is consequently never vintage labeled.

The finos move faster through the solera system, maintaining a more delicate profile, and the olorosos tend to take their time gaining a powerful structure with rich, round textures.

Stylistically speaking, Sherry typically falls into two distinct categories: lighter-bodied finos or dry, fuller-bodied olorosos. All Sherries are initially produced in a dry form and then the olorosos may be sweetened with the concentrated juice of Pedro Ximénez grapes if desired. The fortified nature of Sherry brings the alcohol content into the 15 to 22 percent range in the final bottling, depending on the specific style.

Types of Sherry

In Spain, the vast majority of the local Sherry is produced in the fino style. Deliciously dry and quite delicate, it makes an enthusiastic partner for the abundance of fresh, local shellfish, and tapas.

Overall, Sherry offers a considerable amount of food-pairing versatility, especially given the numerous styles and expressions that emerge from the solera system. Let's look at some specifics:

Fino   Pale colored, with delicate aromas that lean heavily on almonds and apple or pear, with a light body, tangy taste, and very dry nature, this is a fresh, crisp fortified wine with alcohol in the 15 to 17 percent range. Fino pairs particularly well with regional almond and olive staples, an incredible array of local seafood, sausages and dry-cured *jamón* (Iberian ham), and sheep's milk cheese like Manchego. Serve chilled.

Manzanilla   From the coastal town of Sanlúcar de Barrameda, manzanilla is a light-bodied, ultra-dry Sherry that brings a tangy touch with an almost salty quality. Similar to fino, this Sherry style makes a splash when paired with the countless shellfish dishes, seafood soups, and fried fish options of Andalusia. Serve chilled.

Amontillado   Essentially the halfway point between a fino and an oloroso, amontillado Sherries are made as finos and then fortified and run through the solero system again without the cover of flor to allow for a bit of oxidation. The extra step brings a rich, mahogany color; caramel character; and nutty nuances in an off-dry to dry, crisp style. Consider pairing amontillados with smoked seafood, assorted tapas, *jamón*, olives and almonds, and aged cheese. Serve chilled.

Oloroso   The fuller-bodied oloroso Sherries are where things can veer from dry to sweet. The dry styles of oloroso maintain their generous, round textures and are delicious partners for beef, pork, hearty stews, and poultry. The sweeter olorosos contain some of the concentrated sugar of the Pedro Ximénez grape. Rich and flavorful, these fortified wines carry caramel, honey, and nuts on the palate profile and are remarkable for pairing with a variety of desserts. Serve slightly chilled.

Cream Sherry   These Sherries are made when the dry-styled oloroso is pumped up with even more of the Pedro Ximénez grape juice—15 percent. Not surprisingly, cream Sherry offers a creamy, smooth texture with vanilla, warm pumpkin pie spice, and toffeelike character playing on the palate. Cookies, cakes, fresh fruit pies, and tarts top the "must-pair" list for this unique fortified find.

Pedro Ximénez   Spotlighting a very full body, with stunning levels of sweet and a velvety, almost maple syrup–like texture, Pedro Ximénez is fully capable of functioning as dessert itself. Many people prefer to simply top desserts with a dazzling drizzle (think vanilla ice cream) or enjoy with all things chocolate.

PRODUCERS TO TRY

Croft, Emilio Lustau, Hidalgo, Osborne, Pedro Domecq, Sandeman

PORT

Port is Portugal's acclaimed sweet fortified wine made along the dramatic Douro River Valley in the rugged region of northern Portugal. Named after the hilly, coastal city of historic Porto, where Port trading was established, Port has been the world's most sought-after fortified wine for centuries.

The British were the first serious Port customers, looking to source their wine from somewhere other than France during the continuous conflicts of the seventeenth and eighteenth centuries. Portugal was geographically just south and around the Bay of Biscay from Bordeaux, the preferred French wine region for most of England, and was more than willing to start shipping its Douro reds to Great Britain.

The journey north was long and unpredictable, and in an effort to help stabilize the wines on their potentially extended and tumultuous voyage, an increasing percentage of brandy was added to the wine. Mixed in during the fermentation process, brandy brought the conversion of sugar to alcohol to a screeching halt and arrested the wine with considerably higher levels of residual sugar (in the 10 percent range), not to mention alcohol (in the 20 percent range). The result was—and is—a richly fortified wine with dark fruit character, incredible body, and a splendid splash of sweet.

Today, the English influence remains as many of the well-known Port Houses of Porto bear British names: Churchill, Dow's, Warre's, Graham's, Taylor Fladgate, Croft, and many more.

Many regions around the world try their hand at making a Port-style wine—often with decent results—but the real deal can always be identified by the word *Porto* on the label.

Port Styles

Port is made from a blend of predominately red, native grapes, with the top contributors being Touriga Nacional, Touriga Francesca, and Touriga Roriz (same grape as Spain's Tempranillo). Some white wine Ports are crafted locally, typically with decent doses of sweet and lower levels of alcohol compared to their more popular red wine Port cohorts.

Port production basically falls into two distinct categories: the fortified wines that are aged in wood casks and the wines that are aged in bottles. The Ports coming out of wood are generally ready for immediate consumption, while the bottle-aged Ports, of which Vintage Port is the most notable, can take another decade or two to really find their footing.

Port tastes the best when it's served relatively cool—about 60°F to 65°F (16°C to 18°C)—and most can hold themselves together for another few weeks after opening if refrigerated.

There are nearly a dozen different styles of Port. The following variations are the most popular and tend to be the easiest to access.

Ruby Port   Young and relatively simple in style compared to the complexity of the aged and Vintage Ports, Ruby Port spends close to 3 years in wood and consists of a blend of grapes and vintage years. Ruby in color, fruity in flavor, and ultra affordable, Ruby Ports are terrific, easygoing, entry-level ambassadors for the full Port portfolio. Consider pairing the fresh, forward fruit profile with cherry-based desserts, chocolates, and a variety of blue cheeses.

Tawny Port   Fuller bodied, silky textured, amber colored, the Tawny Port style is made with a specific blend of wine from older vintages. The average age of a Tawny Port blend is clearly labeled as *10 year old, 20 year old, 30 year old,* and so on. The price points increase with the average age, as does the inherit complexity and overall intrigue. Creamy caramel and toffee, the concentrated figlike character of dried fruit, spiced nuts, and coffee with a streak of vanilla all play the palate profile of a well-managed Tawny. With the sweet factor ranging from very sweet to semisweet, these fortified wines take well to an array of desserts. Consider chocolate pecan pie, apple tart, pumpkin pie, chocolate soufflé, and caramel treats with the smooth flavors and luscious layers of a well-built Tawny Port.

Late Bottled Vintage Port (LBV)   As the name implies, Late Bottled Vintage Ports are made with grapes all sourced from the same vintage and bottled 4 to 6 years later, after a serious stint in wood. Late Bottled Vintage Ports come packed with dark berry fruit (think black cherry and blackberry) along with a fairly full body and a reasonable price tag. Perfect for pairing with assorted nuts, chocolates, and pastries, LBVs are ready to drink upon release.

Vintage Port   The crème de la crème of the Port world, Vintage Port is only made in the best years, from the best grapes. Showcasing a strategic blend of stellar wines from the same vintage, the iconic Vintage Ports are bottled after only 2 years and then released for patient collectors and consumers to age in their own cellars for at least another decade, if not two. That patience is rewarded with the ambrosial aromas of dark chocolate, espresso, toffee, sweet spice, and dried fruit melded with the refined power and fantastic density of opulent textures and focused flavors on the palate. Certainly dessert on its own, the sweeter side of Vintage Ports brings out the very best in the salty character of Britain's Stilton blue cheese—a classic, enduring pairing that simply can't be beat. Vintage Port demands decanting for aeration as well as sediment and won't store for long after it's opened.

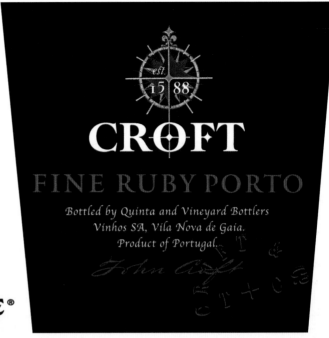

TAYLOR FLADGATE®

20

YEAR OLD TAWNY PORTO

Aged for 20 years in wood

ESTABLISHED IN 1692 • PRODUCT OF PORTUGAL

BOTTLED IN OPORTO BY
TAYLOR FLADGATE & YEATMAN
VINHOS S.A., VILA NOVA DE GAIA

ALC 20% BY VOL   750 ML

⚜ PRODUCERS TO TRY ⚜

Churchill, Cockburn's, Croft, Dow's,
Fonseca, Graham's, Niepoort, Sandeman,
Taylor Fladgate, Warre's

MADEIRA

Madeira is another Portuguese fortified wonder from the volcanic island bearing the same name that's locked in the Atlantic Ocean about 300 miles (483 kilometers) due west of Morocco and 500 miles (805 kilometers) southwest of Portugal. Surrounded by sun and surf, the island of Madeira is known for producing wines that have been well fortified with brandy and allowed to basically bake a bit following fermentation—generally considered a bad thing for wine.

In the case of Madeira, however, the intentional heating of the wine (called *estufagem*) for anywhere from months to many years, depending on the quality and specific style, can produce a big, full-bodied wine with delicious caramelized sugar components wrapped in thick, velvety textures.

The History of Madeira

Madeira was discovered accidentally by trading ships that began to fortify their wines with brandy out of sheer necessity, to stabilize and protect their delicate structure as much as possible. The vessels consistently sailing the southbound trade routes found that their fortified wines were being baked in the hold during the course of the voyage. Instead of being thrown overboard, these "maderized" wines became cherished treasure, offering rich, full-bodied, well-fortified wines often with brown sugar; roasted nuts; rich, creamy, butter-laden qualities; and vanilla latte–like undertones.

Proven to be incredibly resilient to the ship's constant motion, heat, and oxidation, Madeira soon started making the extended journey to the British colonies in North America. It quickly became a favorite drink of America's founding fathers, who toasted significant events like the signing of the Declaration of Independence, George Washington's inauguration, and the establishment of the country's capital in Washington, D.C., with an uplifted glass of Madeira.

Remarkably, bottles of Madeira from the 1800s are still available today, offering a tremendous sip of history and sweet luxury.

Madeira Styles

Madeira is a fortified white wine produced from four primary grapes—Bual, Malmsey, Sercial, and Verdelho—that comes in several different styles ranging from fairly light-bodied to full-bodied and dry to very sweet. The timing of the brandy fortification determines the degree of sweetness. The earlier in fermentation fortification takes place, the sweeter the Madeira is, as yeast is killed off by the brandy, which prevents further conversion of sugar to alcohol and leaves the wine with higher levels of residual sugar (and alcohol).

Because these fortified wines have already experienced high temperatures, oxidation, and significant fortification, they're virtually indestructible and should last for years after they're opened. The drier the style, the cooler it should be served, with the thick, heavy bodies of the sweet Madeiras showing their best at room temperature.

Here are the different styles of Madeira:

Sercial   This is the driest and lightest-bodied style of Madeira, with extraordinary levels of acidity and a rather crisp texture. Almonds, walnuts, citrus, and warm spice are hallmark aromas and work out best as an aperitif, particularly with a medley of roasted nuts, salty olives, or traditional consommé.

Verdelho   This style tends to lean toward the medium-dry side with a bit more body and a decent dose of lively acidity. Verdelho Madeira also pairs well with a variety of aperitifs.

Bual   Bual brings darker, amberlike color components in a fuller body with higher levels of residual sugar and plenty of acidity tamed in silky, rich textures. Rich, nutty character with caramel and toffee qualities makes this a perfect wine for pairing with dessert. Try it with pecan pie or pralines and cream, or tone down the sweets and partner it with foie gras or blue-veined cheese.

Malmsey   The sweetest style of Madeira, with a very full body and intense acidity, Malmsey essentially becomes dessert in a glass with its dynamic notes of exotic cocoa, coffee, brown sugar, caramel and cream, and extraordinary buttered pecan nuances. It's delicious even on its own.

PRODUCERS TO TRY

Barbeito, Blandy's, Broadbent,
D'Oliveira, Leacock's,
Rare Wine Co.

DESSERT WINES

Sweet fortified wines, decadent bottles of late harvest wines, and intensely concentrated ice wines all come together to create a tremendous selection of delicious, dessert-friendly wine options for the savvy sweet tooth.

Dessert wines cover an immense amount of wine-making ground in terms of grape variety, extreme wine-making techniques, and diversity of style and palate approach, but they all share a common denominator—generous levels of residual sugar. The sweet wine journey to elevated levels of residual sugar can take remarkably different routes, from fungus, to frost, all the way to fortification. The many styles of dessert wine are merely a final expression of a very creative wine-growing and wine-making process.

Let's start with the fungus. Late harvest wines often form a symbiotic relationship with botrytis, a fungus that concentrates the grape's sugars by means of functional dehydration. In certain regions, autumn weather conditions—mainly cool, misty mornings and warm afternoons—provide the right mix of moisture and sunshine to encourage botrytis growth and shriveled grapes. Press these prized raisins, and you get liquid gold, albeit in small amounts, which is why late harvest wines come with high price tags and low volumes (often 375-milliliter bottles). The most celebrated (and expensive) late harvest wines are Bordeaux's Sauternes, Germany's Trockenbeerenauslese, and Hungary's Tokaji. All three are botrytis-induced beauties.

Ice wines are an ingenious retort to otherwise uncooperative cold-climate viticulture conditions. Embracing the thrill of the chill, vintners leave berries on the vine long past late harvest until full-on frost sets in, in a gutsy effort to harvest frozen grapes. The frozen grapes are hand-picked and very gently pressed. The super-sweet juice is fermented, bottled (also in half bottles), and priced at a premium. Canada, Germany, and Austria have the market cornered on ice wine (labeled *icewine*—one word—in Canada and *Eiswein* in Germany and Austria).

Finally, many fortified wines are crafted in a sweet, dessert wine style. They provide compelling options as dessert themselves and stellar partners for a range of confections.

LATE HARVEST WINES

Late harvest wines, as the name implies, are made from grapes harvested later in the season. Overripe grapes that have started to shrivel and wrinkle like raisins are brimming with maximum sugar but not a lot of water. Generally, Riesling, Gewürztraminer, Semillon, and Sauvignon Blanc are the go-to grapes for late harvest wines.

These grapes may or may not be affected by botrytis, a fungus that further concentrates the grapes' innate sugars by shriveling the grape skins, quickening water loss, and encouraging the concentration of both the grape's inherent sugar and overall acidity. Fermentation can be tricky with botrytized grapes because the fungus can interfere with the yeast's conversion of sugar to alcohol, often forcing winemakers to intervene to ensure adequate levels of alcohol. Well-known international late harvest wines that rely heavily on botrytis include Bordeaux's exquisite Sauternes, Germany's preeminent Beerenauslese and Trockenbeerenauslese, Alsatian Selection de Grains Nobles, and Hungary's world-class Tokaji.

Plenty of New World and Old World regions take full advantage of botrytis clusters as well as fungus-free, late-hanging grapes that are well concentrated and capable of delivering delicious late harvest wines. However, there is an unmistakable opulence from wines that have been positively influenced by botrytis. Powerful, full-throttle aromas bring highly concentrated whiffs of peaches and cream, ripe apricot, and candied pear wrapped in ongoing honeysuckle, vanilla, and traces of caramel pecan. The massive body, luxurious honeyed textures, and delicate dance between overtly sweet and racy acidity make late harvest dessert wines decadent, ultra-rich, and perfect as dessert alone or partnered with a medley of sweet treats.

Intended to be served chilled to bring out the brightest aromas and fullest flavors, late harvest wines are ideal for pairing with crème brûlée, cheesecake, almond shortbread, lemon poppy seed cake, and fruit-based desserts. Or opt for the contrast of savory sorts like the classic pairing of Sauternes with foie gras or Roquefort blue cheese.

ICE WINES

Genuine ice wines are made from grapes that have thoroughly frozen on the vine—considered the "traditional method" by the elite producers of Canada, Germany, and Austria. Regions that don't reach the chill needed to freeze grapes in the vineyard may have to cut a few corners and pop their grapes into the deep freeze to simulate the conventional process, although often with somewhat diminished exuberance.

Once the grapes are frozen, they're carefully pressed to release the tiny drops of juice carrying the condensed sugar components and natural fruit acids and to separate the water, now frozen. The concentrated juice is then fermented; occasionally aged in oak; and ultimately bottled in small, half bottles due to the limits in supply.

Like many nonfortified dessert wines, ice wines tend to weigh in with lower levels of alcohol, generally in the 7 to 11 percent range. Typically made with white wine grapes in medium- to full-bodied styles, with the intense aromatics of candied fruit, exotic tropical notes, and a fresh floral component, ice wines can bring the intense flavors of bright citrus, ripe apricot, classic marmalade, and poached pear nuances on the ample palate profile.

Riesling and the French American hybrid Vidal Blanc are traditional grape favorites when it comes to ice wines, but plenty of producers around the world are experimenting with different varieties and creating intriguing versions of this famous dessert wine.

The considerably elevated levels of residual sugar—often 25 percent or more—are well balanced by solid levels of lively acidity, which dismisses the syrupy sweet and sticky possibilities and brings an unexpected freshness to the glass.

Ice wines are meant to be served chilled with a variety of fruit-themed confections, creamy custards, and sugar cookies.

APPENDIXES

GLOSSARY

acidity   A component of wine that gives it a zesty, crisp, lively mouthfeel.

aeration   The process of allowing wine to mix with air to soften and open a young wine; most common with tannic red wines.

appellation   A delineated, geographical area of a specific wine-growing region.

aroma   A wine's smell, including influences from the grape itself, fermentation, and barrel aging.

balance   The state achieved when a wine's key components—fruit, sugar, alcohol, acidity, and tannins—are all in harmony with one another.

bodega   A Spanish wine estate.

body   The weight of a wine on the palate, a wine's body is most influenced by its alcohol content. The lighter the alcohol levels (10 percent or less), the lighter the wine's body. Full-bodied wines carry higher levels of alcohol (14 percent or higher).

botrytis   A "friendly" fungus, *Botrytis cinerea* attacks grapes used for making some sweet dessert wines. It concentrates flavors and aromas, resulting in a honeylike textures and flavors. Also called noble rot.

brut   Refers to a dry sparkling wine.

chateau   A French winery estate (often in Bordeaux).

complexity   Perhaps one of the wine world's most overused words, *complexity* refers to the layers of interest that arise from the various aromas, tastes, and textures of a good wine. Multi-dimensional, concentrated character distinguishes a complex wine from a simple, everyday bargain beauty.

decant   Pouring a wine, typically red, into a larger vessel to either open and aerate a young, tannic red or remove sediment from an older bottle of wine.

domaine   A French wine estate (often in Burgundy).

dry   A wine that is not sweet. When fermentation has converted the majority of the sugar to alcohol, the wine has very little residual sugar remaining, rendering it dry.

estate   A winery that grows its own grapes and makes its own wine. The wine is typically referred to as "estate grown" or "estate bottled" wine.

fermentation   The conversion of a grape's sugar to alcohol by the metabolizing action of yeast.

finish   Essentially the "aftertaste" of a wine, the finish is how a wine ends. It's what you taste and sense after you've swallowed a wine. A short finish lasts for seconds, and a long finish could take a minute.

First Growth   Refers to the famous Classification of 1855, designating the top five chateaus in Bordeaux.

fortified wine   A wine that has additional alcohol, often brandy, mixed in.

fruity   A wine that has heavy fruit-based aromas and flavors, at times referring to specific fruits.

ice wine   A dessert-style wine made from the concentrated juice of frozen grapes.

lees   The dead or "spent" yeast and other grape particles that collect on the bottom of a fermentation vessel.

legs   The term used to describe the streaks of wine left on the inside of the glass after the wine has been swirled; can indicate higher levels of alcohol or a fuller-bodied wine, but not always.

maceration   A process wherein the skins from red wine grapes are kept in contact with the fermenting grape juice to add color, tannin, and flavor to the wine.

malolactic fermentation   Often abbreviated as *malo*, this is the chemical conversion malic acid (think green apples) to lactic acid (think milk), creating a softer palate impression.

minerality   A wine description that refers to the aromas or flavors of minerals, often flint, slate, gravel, or chalk, present in a wine.

mouthfeel   How a wine's components—alcohol, sugar, tannin, and acidity—feel on the palate. Are they heavy, creamy, crisp, lean, or dry?

must   The mix of grape juice and solids (seeds, stems, and skins) from freshly pressed or crushed grapes.

New World   The wine-making countries of North America, South America, New Zealand, Australia, and South Africa.

noble rot   *See* botrytis.

oak   The preferred wood for aging wines. Oak impacts a wine's aromas (vanilla, spice, butter, smoke) and textures (creamy, rich) and increases the tannin content. French, American, and Eastern European varieties of oak are the most popular for making barrels.

Old World   The dominant wine-making countries of Europe.

organically grown wine   Wine made from grapes that have been grown without the use of synthetic fertilizers, pesticides, herbicides, or fungicides. They might have added sulfites for stabilization.

palate   Refers to the feel and flavor of a wine in the mouth.

phylloxera   A root-eating louse that devastated vineyards of major wine-growing regions in the 1800s and early 1900s.

residual sugar (RS)   The sugar that remains after fermentation has stopped.

sommelier   A person responsible for the wine service, and often selection, in a restaurant.

still wine   A table wine that has not undergone second fermentation (for bubbles) or been fortified.

tannin   The dry, pucker power in a wine. It's the astringent, sometimes bitter texture in a wine that comes from the grape's seeds, skins, and stems and is most notable in red wines. Tannins offer the wine structure and aging potential.

terroir   The place the grape was grown and all the factors that influence the vine. Sun, soil components, climate, topography, geography, irrigation and drainage, and weather patterns all play into a grape's terroir.

texture   The tactile feel of a wine in your mouth. The texture could be silky, velvety, creamy, harsh, or abrasive, among other traits.

unfiltered   A wine that has not undergone filtration to remove small sediments. Some believe filtration may strip a small portion of the innate aromas and flavors from the final wine.

unfined   A wine that has not been exposed to common fining agents, like egg whites, bentonite, or gelatin, that capture and remove tiny particulates from the wine.

varietal   A specific type of grape variety such as Chardonnay, Cabernet Sauvignon, or Pinot Noir.

vinification   The process of making wine from grapes.

vintage   The year a wine's grapes were harvested.

vintner   A winemaker.

viticulture   The science or study of cultivating grapes.

WINE Q&A

With so many wine-producing countries, grape varieties, wine-making styles, and vintage selections, questions often arise. Let's take a moment to answer some of the most commonly asked questions.

What's the Best Bang for My Wine Buck?

If you're looking to get to know wines from regions all over the world without spending a lot of money (most are under $20/€15.35), the following producers bottle solid, value-driven wines along with upper-end splurges. Most are well distributed, offering a slew of entry-level delights for curious consumers.

Alamos	Chateau Ste. Michelle	Ken Forrester	Robert Mondavi
Alois Lageder	Clos du Bois	Kim Crawford	Rosemount
Antigal	Columbia Crest	King Estate	Ruffino
Antinori	Concho y Toro	Kunde	Rutherford Ranch
Banfi	Cono Sur	La Crema	Segura Viudas
Beaulieu Vineyard	d'Arenberg	Lapostolle	Simi
Beringer	Domaine Faiveley	Lindeman's	St. Supéry
Bodegas Montecillo	Dr. Loosen	Louis Jadot	Susana Balbo
Bodegas Muga	Dry Creek	Louis M. Martini	Trapiche
Bogle	Freixenet	Merryvale	Trimbach
Chalone	Frescobaldi	Mionetto	Veramonte
Château de Macard	Hess	Nino Franco	Villa Maria
Château de Reignac	Hogue	Penfolds	Viña Eguía
Château Greysac	J. Lohr	Peter Lehmann	Yalumba
Château Meyney	Jaboulet	Rancho Zabaco	Zaca Mesa
Chateau St. Jean	Jospef Leitz	Ravenswood	

What Are the Best Bets for High-End Red Wines?

When it's time to splurge, either for yourself or as a special gift, look to the heavy-hitting reds of California, Italy, and France. These are just starting suggestions; as you get to know the world of higher-end reds, you'll find others you like.

In California, try the following:

Cakebread Cellars	Joseph Phelps	Robert Mondavi	Stags' Leap
Caymus	Ladera	Shafer Vineyards	
Groth	Merryvale	Silver Oak	

From France, consider these:

Chateau Duhart-Milon Rothschild	Château Haut-Bailly	Château Palmer	Château Smith Haut Lafitte
	Château Montrose	Château Poujeaux	

Here are Italy's big boys:

Antinori	Frescobaldi	Ornellaia	Tenute Silvio Nardi
Ceretto	Gaja	Sassicaia	Valdicava

What About Screw Caps?

Screw caps are great closures that keep bottles fresh and prevent cork taint. Plus, you don't need to be concerned with storing wines in low humidity environments or worrying about corks shrinking and allowing air to oxidize. More wine-producing regions are leaning toward screw cap closures, with Australia and New Zealand blazing the trail.

What's the World's Most Versatile Food-Pairing Wine?

Bubbles, sparkling wines, and Champagne are the most food-friendly, versatile wines around for pairing. The high acidity creates plenty of perfect pairing partners, and the bubbles bring on fun, festive moods. Try them with a variety of hors d'oeuvres, exotic dishes, spicy fare, and pastas.

What Are the Best Temperatures for Serving Wine?

For most whites, aim for 45°F to 55°F (7°C to 12°C), and for reds, opt for 60°F to 65°F (16°C to 18°C). Keep sparkling wines a little cooler, right around the 45°F (7°C) mark, and rosés at 45°F (10°C).

What Causes Wine Headaches?

Most people blame sulfites for headaches, but dehydration also plays a role. Science has been pretty inconclusive on firmly laying the blame on sulfites, histamines, or other innate wine ingredients, although excessive additives could certainly affect the metabolism of wine and how it impacts an individual's physiology.

Are Pricier Bottles Worth It?

It depends on the price and the bottle. Many high-end wines pay extra for single-vineyard grapes from some of the region's best vines, and they choose to age their wines in 100 percent new oak (a huge expense) and take the time (and space) to bottle-age their wines prior to release. When all these factors come together, they demand a higher price tag.

Other bottles may raise the price, simply based on a name or label recognition. These might suit your palate perfectly well, or they may disappoint based on their price expectations.

Plenty of lower-priced bottles of wine could beat a costlier bottle in a blind taste test based on palate appeal, approachability, and pairing versatility. It just depends on the region, producer, and your personal taste preferences.

How Long Will Wine Keep After It's Opened?

As soon as the cork is popped from the bottle, the wine inside is on its way toward oxidation. You can push a wine for another 2 or 3 days using refrigeration, removing or displacing oxygen before reinserting a cork, or using a smaller half-bottle to decrease total oxygen exposure, but that opened bottle of wine will only be good for a few days after opening.

RESOURCES

To help further your wine education, check out these books and websites.

Recommended Reading

Clarke, Oz, and Margaret Rand. *Grapes and Wine: A Comprehensive Guide to Varieties and Flavours.* New York: Sterling Epicure, 2010.

DeSimone, Mike, and Jeff Jenssen. *Wines of the Southern Hemisphere: The Complete Guide.* New York: Sterling Epicure, 2012.

Dornenburg, Andrew, and Karen Page. *What to Drink with What You Eat.* New York: Bulfinch Press, 2006.

Goldberg, Howard G. *The New York Times Book of Wine: More Than 30 Years of Vintage Writing.* New York: Sterling Epicure, 2012.

Johnson, Hugh, and Jancis Robinson. *The World Atlas of Wine.* London: Mitchell Beazley, 2007.

MacNeil, Karen. *The Wine Bible.* New York: Workman, 2001.

Oldman, Mike. *Oldman's Guide to Outsmarting Wine: 108 Ingenious Shortcuts to Navigate the World of Wine with Confidence and Style.* New York: Penguin, 2004.

Robinson, Andrea Immer. *Great Wine Made Simple: Straight Talk from a Master Sommelier.* New York: Broadway Books, 2005.

Robinson, Jancis. *The Oxford Companion to Wine, Third Edition.* New York: Oxford University Press, 2006.

Shafer, Doug, with Andy Demsky. *A Vineyard in Napa.* Berkeley: University of California Press, 2012.

Zraly, Kevin. *Kevin Zraly's Windows on the World Complete Wine Course.* New York: Sterling Epicure, 2012.

Wine Websites

Wine websites are very prevalent these days, but some seem to have serious staying power based on fresh wine takes, interesting conversations, easy education for budding wine enthusiasts, and straightforward wine-buying opportunities and advice:

1 Wine Dude
1winedude.com

About.com Wine
wine.about.com

Burghound.com
burghound.com

The Daily Meal
thedailymeal.com

Decanter
decanter.com

Dr. Vino
drvino.com

eRobertParker.com
erobertparker.com

Fermentation: The Daily Wine Blog
fermentationwineblog.com

***Gourmet* magazine**
gourmet.com

JancisRobinson.com
jancisrobinson.com

K&L Wine Merchants
klwines.com

New Zealand Wine
nzwine.com

Oregon Wine
oregonwine.org

Snooth
snooth.com

Steve Heimoff
steveheimoff.com

Vinography: A Wine Blog
vinography.com

Washington State Wine Commission
washingtonwine.org

Wine.com
wine.com

***Wine and Spirits* magazine**
wineandspiritsmagazine.com

Wine Business
winebusiness.com

Wine Enthusiast
wineenthusiast.com

Wine Institute
wineinstitute.org

Wine Spectator
winespectator.com

Wines of Chile
winesofchile.org

Wines of Germany
germanwineusa.com

Wines of Portugal
winesofportugal.info

Wines of Spain
winesfromspain.com

INDEX

C

D

G

X–Y–Z

PHOTO CREDITS